The Green Vision of Henry Ford and
George Washington Carver

ALSO BY QUENTIN R. SKRABEC, JR.
AND FROM MCFARLAND

The Carnegie Boys: The Lieutenants of Andrew Carnegie That Changed America (2012)

Edward Drummond Libbey, American Glassmaker (2011)

Henry Clay Frick: The Life of the Perfect Capitalist (2010)

H.J. Heinz: A Biography (2009)

The Metallurgic Age: The Victorian Flowering of Invention and Industrial Science (2006)

The Green Vision of Henry Ford and George Washington Carver

Two Collaborators in the Cause of Clean Industry

QUENTIN R. SKRABEC, JR.

McFarland & Company, Inc., Publishers
Jefferson, North Carolina, and London

LIBRARY OF CONGRESS CATALOGUING-IN-PUBLICATION DATA

Skrabec, Quentin R., Jr.
The green vision of Henry Ford and George Washington Carver : two collaborators in the cause of clean industry / Quentin R. Skrabec, Jr.
    p.   cm.
Includes bibliographical references and index.

ISBN 978-0-7864-6982-6
softcover : acid free paper ∞

1. Ford, Henry, 1863–1947. 2. Carver, George Washington, 1864?–1943. 3. Manufacturing industries — Environmental aspects — United States — History — 20th century. 4. Industries — Environmental aspects — United States — History — 20th century. 5. Chemurgy — History — 20th century. I. Title.
HD9710.U52F66635 2013   338.973'07 — dc23   2013005775

BRITISH LIBRARY CATALOGUING DATA ARE AVAILABLE

© 2013 Quentin R. Skrabec, Jr. All rights reserved

No part of this book may be reproduced or transmitted in any form or by any means, electronic or mechanical, including photocopying or recording, or by any information storage and retrieval system, without permission in writing from the publisher.

On the cover: *inset* George Washington Carver, Henry Ford (Library of Congress); *background, left* peanut plant (Clipart.com); *right* interior of Ford Motor Company Long Beach Assembly Plant (Library of Congress)

Manufactured in the United States of America

*McFarland & Company, Inc., Publishers
Box 611, Jefferson, North Carolina 28640
www.mcfarlandpub.com*

To both my mothers — Dory and Our Lady of Grace
And my high school science teacher
Marylouise Gass (Sister Barbara Francis)

Acknowledgments

Once again I would like to thank the staff of the Benson Research Center at The Henry Ford for their excellent help. In particular, Peter Kalinski was invaluable in my research. They have been at the core of building my literary pantheon, which contains many honored in Greenfield Village such as H.J. Heinz, Edward Drummond Libbey, William McKinley, George Westinghouse, and William McGuffey. Furthermore, Greenfield Village and The Henry Ford gave endless hours of inspiration. Another new source of research was found in the Austin Curtis papers of the Bentley Historical Library at the University of Michigan. These papers of Carver's assistant offered a source of material on Carver not previously used. At the Bentley Historical Library, I would like to thank Malgosia Myc and all the staff of this outstanding library. Finally, thanks to a great friend and constant research companion, Julian N. Robur.

Table of Contents

Acknowledgments   vi
Preface   1
Introduction   5

1. Declaration of Dependence on the Soil   11
2. Common Soil   21
3. Finding Their Mission   29
4. In Search of Destiny at the Fair   36
5. Tuskegee and Detroit   43
6. The Model T and the Jesup Wagon   64
7. The Industrialist and the Professor: Capitalism and Agriculture   77
8. Massive Assembly, Peanuts, and Aircraft   93
9. A Better Life — Industrial and Agricultural Utopias   106
10. The Sages of Dearborn and Tuskegee   119
11. Green Supply Chain for America's Industry — Village Industries   127
12. Chemurgy: A New "Political" Science   140
13. The Meeting of Ford and Carver   154
14. The New South, Soybeans, and Final Dreams   163
15. Home and Industrial Economics — Lean Manufacture and Household Savings   171
16. A Shared View of the Education of Our Youth   178
17. The Last of the Victorians   186
18. Continuing Legacy   193

Timeline   199
Chapter Notes   201
Bibliography   207
Index   209

Preface

I have spent the last twenty years researching and writing, building a literary pantheon of American capitalists and businesspersons. This has been a true passion, but researching George Washington Carver and Henry Ford has been an added pleasure. It reignited my own childhood loves of chemistry, geology, and science. As a materials engineer, I found a closeness with Carver's work and Ford's engineering. It returned me to four years of personal heaven at the University of Michigan. Clearly, both these men deserve a seat in any pantheon of American capitalism. They represent the twin gods of creativity and innovation. While Carver held no patents, he deserves a place with George Westinghouse, Thomas Edison, H. J. Heinz, and others in the inventive wing of the pantheon.

As an industrial biographer with a mission to build a pantheon of American greats, I choose my biographies based on both business success and virtue in hopes of truly creating models for future business leaders. Ford and Carver certainly meet the criteria. Still, all men come with faults. Henry Ford brought his anti-Semitism, but on balance, he contributed much to the advance of blacks, the disabled, poor youth, and society as a whole. Few men are saints and all saints are men. George Washington Carver, however, was unusual; only in George Westinghouse do I see virtue on Carver's level. Carver faced long odds and much bias in his jump to an American pantheon of industrial greatness. He has no equal in changing America just by his life with no other activism needed.

The book is biographical but is not meant to be a biography. The early lives and careers of both men are developed in relationship to their ultimate friendship and merger of passions. The book focuses on the points in their lives that made them part of the chemurgical, green movement in the last years of their lives. In this respect, the reader of Ford and Carver biographies will find a different perspective and new insights into their lives. Their lives clearly led these men on different paths to arrive at the symbiotic relationship of agriculture, industry, and the environment.

Another unique part of this literary exploration has been the study of the rebirth of chemurgy today as the green science movement. The book starts with Henry Ford's chemurgical conference at his reconstruction of Independence Hall in Dearborn, Michigan. It was here that Henry Ford would first meet George Washington Carver. I was at a 2007 kickoff of Mitt Romney's failed presidential campaign in the same building when he recalled that signing. The theme that day was the return of American industry and exceptionalism, which is close to my heart and personal mission. The theme of manufacturing, agricultural, and energy independence had been why Ford built a replica of

Preface

Independence Hall in the heart of Detroit's manufacturing plants. This theme of Ford's is often overlooked by the tourist who visits Ford's great museum.

It reminded me of the vision of industrial and national independence that Henry Ford wanted to establish. His exact replica of Independence Hall at the center of the automotive industry stood for this industrial vision. In many ways, the diverse movements today of the different political parties, unions, management, alternative energy, exploration of fossil fuels and so on are rooted in the same goal of American independence. It is my hope that my literary pantheon will lead future innovators, scientists, manufacturers, politicians, and business leaders to Ford's vision of industrial and agricultural independence for inspiration.

The Ford Museum and Greenfield Village (known more formally as the Edison Institute) is where the flame of American industry still burns. It is a true shrine to America's best businessmen, scientists, and engineers. Like our Founding Fathers and their spirits still haunting Philadelphia's Independence Hall, here at The Henry Ford are the spirits of America's founding capitalists and industrialists. The artifacts of Edison, Ford, Westinghouse, Heinz, the Wright Brothers, and many others are here to inspire. When my studies at the University of Michigan's engineering school overwhelmed me, I came here to renew my spirit and remember my goals. It is here that I came to start the first pages of this book and would return monthly to search not only facts, but also the very spirits of Henry Ford and George Washington Carver. It is these spirits that I wanted to capture more than anything. In fact, many Ford biographers have stated that Henry Ford was too complex to fully understand. For me, Greenfield Village is the heart and soul of Henry Ford. It tells his life, defines his friends, paints a picture of his beliefs, and immortalizes his vision of a future America. It is here that Henry Ford still walks the streets.

In high school and college, I came to know the roots of Ford's village industry network. In high school, I went to a social function at the Tecumseh Community Center, which had been a Ford water-powered factory. Similarly, my fraternity, Theta Chi, at the University of Michigan, often rented the old Ford factory at Saline for parties. While mostly preoccupied with the partying, I was also fascinated by the remnants of the old factories. At the time I knew nothing of Ford's village industries. That so many farms in the area are tied to Henry Ford also puzzled me. I was getting pieces of a story of a very different Henry Ford. There were lots of stories, too, about Henry Ford and the local farmers. I had no idea that Ford had been the reason for Lenawee County where I went to high school. This book contains the roots of much of that early interest of mine and my days in southern Michigan.

Another goal was to look at the beginning of a new manufacturing, energy, and agricultural revolution. Many have credited Toyota with the new approach of lean manufacturing, even though Toyota had credited Ford. The issue was not in a new methodology of Toyota's manufacturing system, but in its adherence to Ford's original principles. America's manufacturing decline is rooted not in the failure of Ford's mass-production system, but in industry's deviation from it. Our latest look at new energy sources such as ethanol and other renewable fuels is consistent with the early work of Carver and Ford, who pioneered these fuels. Both men had argued for the elimination of polluting gasoline. They preached recycling and conservation, so badly needed today, to a world of plenty. The science of chemurgy that they founded is a science for today. Studying Ford and Carver is a must for lean and green manufacturers. It can bring together environmentalists and industrialists.

This book goes far beyond history and biography; it opens up a vision of industry, agriculture, and the environment working in harmony. Many of Ford and Carver's failures are today's successes. Many of the solutions for America are stored in the archives in The Henry Ford and the Tuskegee Institute. They need to be studied once again as Taciihi Ohno of Toyota did to bring a second revolution to industry based on the principles of Henry Ford. As an operations manager turned professor, I've found that the future can be found in the libraries of the past.

Introduction

> "I, for one, am in awe of Ford's greatness. For example, on the issue of standardization and the nature of waste in business, Ford's perception of things was orthodox and universal."
> — Taiichi Ohno, father of the Toyota manufacturing system (lean manufacture)

Few probably know the shared vision of a green American industry and the unusual friendship of Carver and Ford. They saw the American farm as the core of American industry. They were more than friends; they pioneered a movement to fuel industry from agriculture. They believed in biofuels, clean alternative energy sources, recycling, conservation, new uses for plants, industrial materials from agriculture, and smokeless rural factories. Both Ford and Carver took to the laboratory to study organic chemistry, botany, and industrial chemistry. They called this mix of industrial and agricultural sciences "chemurgy." They foresaw a new class of plastic materials made from soybeans, peanuts, and sweet potatoes. Both men developed recipe books to expand the use of underutilized vegetables, fruits, and plants. Both looked at cow-processed milk as an inefficient route to make milk from plants. They argued for biofuels and various petroleum-based fuels. They led the search for alternative plant sources for rubber and other industrial materials versus oil-based dirty chemical conversion plants for synthetic rubber and other materials. They were pioneers in the field of composite materials. Finally, Ford and Carver believed that rural education was the heart of America's creativity and ingenuity.

Of course, both men had a full career before they would meet. Both were world renowned — Ford as the great capitalist and designer of the Model T and mass production, and Carver as the great black scientist of the Tuskegee Institute and the man who took peanuts and sweet potatoes from novelties to two of the world's major crops. Both men were Victorian "men for all seasons" with what seemed to be endless projects and new areas of interest. Both men had deep roots on the American farm.

They pulled a full circle of friends into this new movement that included environmentalist John Burroughs, botanist Luther Burbank, inventor Thomas Edison, and industrialist Harvey Firestone. They even created a true branch of science, chemurgy, to unite industry and agriculture. Their 1935 chemurgy conference of 300 scientists explored ways to use plants to feed the world, to produce industrial materials and clean fuels, and even explore industrial plant hybrids. Chemurgy proclaimed ethanol as the future fuel of the world at this 1935 conference. Both men went even further, envisioning a symbiotic alliance of industry, education, and agriculture represented by the friendship of Ford and Carver. Both believed that complete dependence on American agriculture would reduce wars.

Ford, the great industrialist, hoped he could somehow create a Jeffersonian agrarian society based on decentralized industry. Ford loved his farm in Dearborn and disliked the industrial urban communities, which he himself had played a major part in creating. Ford hated the pollution of the gasoline engine and saw the fumes as dangerous to health. He argued his whole life for alcohol-based fuels. Ford looked to build utopian combinations of industry, education, and agriculture. He befriended environmentalists, including John Burroughs, and botanist Luther Burbank to help promote a green vision for America. Ford proposed future agrarian cities, factories, and agricultural plantations. Ford even moved to rural, water-powered factories to supply parts to his assembly lines. These rural factories were a network of river-village communities throughout Michigan. These village factories enhanced the rural communities. Ford used waterpower and part-time farm labor to manufacture electrical and mechanical parts for his automobiles. He created over thirty of these village factories. They were often restored gristmills of a previous age. He manned these plants with farmwomen and part-time farmers. Today these waterpowered factories are often community centers in Michigan towns such as Ypsilanti, Dundee, Plymouth, Northville, Saline, and Willow Run that grew up around them.

Similarly, Carver argued for self-sufficiency of the farmer. Carver also saw a link between agriculture and industry. He believed that the black southern farmer could not only live off his land but profit from the industrial uses of crops. To promote this link, Carver researched uses for productive plants such as peanuts, soybeans, sweet potatoes, and others. Like Ford, Carver had a passion for conservation and lean farming. Like Ford, Carver believed that this green basis for America was rooted in education. Like Ford, Carver believed that the strength of America was in its rural value system. Their views, vision, and dreams were clearly cut from the same cloth.

While Ford and Carver would not meet until the last years of their lives, the dreams of both men continued to converge throughout their lives. Both Carver and Ford foresaw an industrialization of the south. They fought hard for this farm-friendly industrialization but never fully saw their dreams come true. Ford with his Muscle Shoals, Alabama, project in the river valleys of Tennessee and Alabama foresaw a hydroelectric system that would be the "Niagara of the South." Like Carver, Ford foresaw a merger of industry, agriculture, and the environment in these southern valleys. Ford proposed a seventy-mile corridor of clean power plants, fertilizer plants, aluminum plants, smokeless steel mills, clean coal mines, cotton fields, and vegetable farms. Had Congress approved the project, Muscle Shoals would today be the largest industrial city in the south. Carver argued for the same alliance of industry and agriculture. He struggled for decades to bring industrial factories to the South. Their visions were bigger than their achievements; but today once again, the nation looks to the industrial and agricultural efficiency and linkage that creates wealth and eliminates waste of our natural resources.

Ford's smokeless green factory vision would require the genius of his friend and partner in the movement, George Washington Carver. Ford used many of Carver's innovations to utilize corn, cotton, linseed oil, wool, flax, soybeans, sugarcane, and farm animals in

Introduction

the production of cars. Ford and Carver would ultimately set up a research lab to use farm products in car production. They developed prototype cars made out of soybean plastic, with vegetable oil for lubrication, ethanol for fuel, and rubber from weeds for tires, hoses, and fuel tanks. Ford and Carver would work together at their Georgia experimental farm to expand the use of plants in manufactured car parts. In 1933 Ford premiered a soybean green car at the World's Fair; and in 1943, he produced a "farm"-built green car.

Ford highlighted this "industrialized barn" approach to manufacturing at the 1933 World's Fair. He would spend over thirty years implementing his vision of a united industry and agriculture. He even created an outdoor museum, Greenfield Village, dedicated to his vision. It is at Greenfield that reconstructed factories and farms stand together. Ford reassembled the early home of Carver, the office of Luther Burbank, and Edison's laboratories to illustrate the link between farm and industry. Ford's dream was never fully realized, but once again, America is looking to a more friendly combination of agriculture and industry. Ford even tried to take his principles of industrial efficiency to the agriculture in his own South American rubber plantations.

It is truly unfortunate that these men did not meet until they were in their seventies. Henry Ford saw George Washington Carver as the Edison of agriculture and a true genius. These two geniuses were themselves the last of a breed of Victorian scientists, when they created the new science of chemurgy. Both of these men had changed the world; but in their seventies, they were ridiculed by a new breed of university scientists and engineers. But neither Ford nor Carver saw their usefulness as over. Near the end of their lives they would be united in a Dearborn laboratory experimenting with sweet potatoes, milkweed, soybeans, and other plants in search of a new source to make rubber to address the shortages of World War II. Henry Ford even built a special guest cabin for Professor Carver near his Greenfield Historic Village.[1] America has known only a handful of geniuses on the level of Carver and Ford. Today, many still question the final visions of these geniuses due to various biases.

Ford was a complex man with many idiosyncrasies, but he can be attributed with his green initiatives and his desegregation of the automotive industry. Carver's experimental and even spiritual approach raised doubts in the scientific academic community. Still, Ford and Carver were drawn together in a final vision of green industry. American industry might have been much different if these two geniuses had gotten together earlier in life. Carver and Ford had a shared circle of friends based on the chemurgical (green) movement. It's a great story of what has been called "uncommon friends." Ford and Carver's circle of friends included Thomas Edison, Harvey Firestone, Luther Burbank, John Burroughs, many secretaries of agriculture and commerce, and presidents Hebert Hoover, William Taft, and Warren Harding.

Industry, big oil, university scientists, and New Deal farm policies would put down this final green revolution of Ford and Carver's, but it has resurfaced today. We are once again moving back to the future. Ford's idea of a lightweight car of high-strength steel and plastics driven by a four-cylinder engine fueled by an ethanol and gasoline mix is now taking to the roads. Ford and Carver's synthetic rubber is in all of our tires. The idea of agriculture and industry functioning in a smokeless environment is making progress once again. Carver's endless uses for plants and minerals are again replacing our dependence on oil. Both Ford and Carver had argued for ethanol and bio fuels that are today finding their way to fueling our cars. The plastic and plant-based composite materials that Carver and Ford pioneered are finding new uses today.

Besides science, both men united in resisting black segregation and racism. The legacy of black rights is in both their personal legacies. Ford was the first to hire blacks in high-paying industrial jobs and bring them into the skilled trades. Carver changed the perception of blacks' ability in the sciences, yet was criticized for his lack of militancy in the early civil rights movement. Clearly both Ford and Carver were great scientists and engineers, yet both men continue to come under attack from university historians. Carver's criticism is more focused on his scientific methodology and the idea that his myth had social benefit.[2] In Carver's case some have identified black groups who would have an interest in making him a myth. The facts of his life show a much different story of opposition to his accomplishments. Ford is often attacked for his support of Nazi Germany in the 1930s, which to some degree is deserved, but great men are just men. They come with weaknesses, and even our saints are flawed. Ford clearly used his position to create myths, but his accomplishments are undeniable.

Both men were a product of their times. Their Victorian methodology seems out of place today, but it allowed for the evolution of human endeavors and science. Carver's simple experiments and Ford's wooden models seem prehistoric in an age of virtual manufacturing and scientific theory. They believed in lab-based experiments, which led to the development of a new science of chemurgy. Chemurgy combined chemistry, botany, and biology with a bit of home economics. It was pure Victorian in its roots when science was moving to a different level. Some of this criticism for both men is rooted in the chemurgical movement itself, which for political reasons was framed as naive and nonscientific. It is interesting that much of what is in vogue with the green movement of today was looked down on in the science of the 1930s.

Ford and Carver at times seemed to be alone against the science of the day. The attacks drew the two together. Ford and Carver formed a close friendship in the last years of their lives. Ford not only brought Carver and his assistant, Austin Curtis, to Dearborn to do research with soybeans but also added Carver's home to his industrial complex of Greenfield Village. Ford built a lab and an experimental farm in Georgia for Carver's use. When Carver had problems walking stairs, Ford had an elevator installed at Tuskegee. Even more intimate was a gift of a cup and saucer from Ford's mother's tea set brought from Ireland. Carver would say in a letter to Ford, "The greatest gift I have ever received from a mortal man is the time I met you for the first time at Dearborn. I was thrilled and inspired as never before. I have been able to work better, you seem to be ever present with me in my investigations."[3]

Both men were frugal and thrifty, and both carried these simple Victorian principles into their careers. They represented the earliest conservationist and green movement. Waste elimination was the cornerstone of all their projects, advancing both lean industry and lean agriculture. Recycling and by-products became a passion to these two. Both

Introduction

promoted these lean principles from the home and school to the workplace. They agreed that gardening, conservation, and recycling should be part of the school curriculum. Ford believed his employees should have home gardens, as did Carver with his farmers. Both saw home economics as a true science. They went further in suggesting a disciplined approach to education of the youth with history and American independence and exceptionalism at the core. Ford built schools throughout the country based on the principles of Carver's Tuskegee Institute. Their views were often on the cutting edge of science and education, reflecting the greatest of America.

The fact is that Ford, Carver, Edison, and many Victorian scientists and engineers are in a class of their own. Genius is always relative to an era. More recently, for the 100th anniversary of the Model T, Ford Motor Company used today's engineers to build replicas. They quickly found that their knowledge would not be used in the past. The Model T was built not only differently than today's engineering but also was built for a different consumer and culture. Ford's Model T and assembly process were deeply rooted in the culture of the times. Similarly, what seems so simple today for agriculture was far different then. Farmers in the south lacked basic education. Carver was an educator for a population that couldn't read the simplest of his bulletins. He was a scientist for former slaves and poor white farmers. Carver had to use visual education and take it to the farmers via a school wagon. Both Ford and Carver had revolutionary visions for a much different society. Both men, through their work, brought equality and freedom to America. They believed that industry, education, and agriculture were the result of as well as the engine for, democracy. Maybe just as important (and unique) was that Carver and Ford saw a harmony among the green movement, education, and industry.

Ford's development of mass production, the Model T, lean manufacturing, and the spread of American history are well known. Carver's achievements in agriculture, chemistry, science, and race relations were less known than those of Ford; however, Carver would find recognition. Carver's naming to the National Agricultural Society in 1916 and his election to the Royal Society of Arts in Great Britain in 1917 had established him as a scientist. Henry Ford would immortalize Carver's achievements, starting with his home being reproduced in Greenfield Village. Ford would say that Carver replaced Edison as the world's greatest scientist. Carver's recognition continued long after his death. Congress created a national monument to him in 1943. The navy named a cargo ship and submarine after him. In 1947 a United States postage stamp was issued with his face on it. Many college campuses have been named for him. Hollywood has made two movies about him.

The ideas of Ford and Carver in conservation, recycling, waste elimination, and education are timeless and a continuing legacy. The U.S. government called on both these men in the times of shortage during both world wars. After Ford's death, American auto manufacturing and industry returned to wasteful production based on economies of scale, which covered the cost of waste. Ford and Carver's visions continue to resurface. After World War II, Toyota came to Ford to study his system and became followers of Ford's approach to waste management. The father of the revolutionary lean manufacturing of Toyota, Taiichi Ohno, studied Ford's writings and based his system on them. Once again, Ford's ideas would revolutionize industry in the 1990s with lean manufacturing.

Introduction

Carver and Ford were prophets in their own right. Today's recycling programs are based on that of Carver and Ford. Even things like the recycling of motor oils were foreshadowed by the work of George Washington Carver, who used the motor oil in paint production. Carver's and Ford's plant substitutes for meat and milk are world-market staples today. Throughout their lives, Ford and Carver fought for an alcohol/gasoline mix to fuel our cars. Alcohol-based fuels are once again fueling our cars. Ford argued for green energy sources long before today's environmentalists. Ford wanted research in clean coal and even the expanded use of waterpower and wind power. Ford's and Carver's writings are beyond visionary; they are handbooks for the future. The dominant core belief of both men was that American independence in the long run was based on its dependence on American agriculture for all its needs.

1

Declaration of Dependence on the Soil

"I agree with everything he *[Carver]* thinks and he thinks the same way I do."
— Henry Ford

More than 300 delegates from all over the nation made their way to Independence Hall to sign what was to be a revolutionary document. This Independence Hall, however, was in Dearborn, Michigan, not Philadelphia; and the year was 1935, not 1776; and it was May 7, not July 4. Henry Ford had only a few years earlier offered to buy, take apart, and reconstruct the real Independence Hall in Dearborn. When Philadelphia refused his offer, he built an exact replica of Independence Hall at his industrial museum, Greenfield Village. Ford assured exact detail in the replica, including the recasting of the Liberty Bell from its original mold in England. These delegates were not the first group of famous individuals to gather here. In 1929, this Independence Hall had opened to celebrate the Golden Jubilee of Edison's lightbulb. That dinner in October of 1929 had featured guests like President Herbert Hoover, Thomas Edison, Madame Curie, Will Rogers, Harvey Firestone, Oliver Wright, Charles Schwab, and many others. It included a live radio address from Albert Einstein. Days later the nation would be plunged into a great economic depression. The nation was feeling the impact of that 1929 on this spring day in May 1935. This conference was bringing together not only great scientists and industrialists, but America's greatest dreamers. Greenfield Village was the ideal location for dreamers. Ford's Greenfield Village was a beacon of hope that American creativity could change the world.

Ford had planned every detail of the conference, with motorcades moving attendees from nearby Ford-built Dearborn Inn and the Cadillac Hotel in Detroit. The meetings were held throughout the village, with the larger meetings at the Edison Institute's main building and the Martha Mary Chapel in the center of the village. The village was a tribute to Edison, Ford, and others. Edison's New Jersey labs and research complex had been taken down piece by piece and shipped to Dearborn to be rebuilt. It was Ford's tribute to American industry and agriculture, which was represented by his own eclectic mix of rebuilt factories and farms. It included Ford's rebuilt farm, stagecoach hotels, William McGuffey's (the educator of great Americans) home and first school, Noah Webster's home from New England, and Edison's New Jersey compound of laboratories. The delegates would get a chance to visit school students in Ford's boyhood school and an early McGuffey school. Ford was also running his own grade school and high school system out of the village to train the next generation of great Americans. He was determined to reinstitute the use of the *McGuffey Readers* and their value system back into American education.

These visiting 300 scientists, industrialists, religious leaders, farm leaders, and government leaders came to proclaim a dependence on the regenerative powers of American agriculture and a new future for American industry. Most were traveling by train to Dearborn, Michigan, to be greeted by Henry Ford. They had come to declare the nation's dependence on its agriculture. That dependency included the nation's industrial might. Ford's conference would be highlighted with a ceremonial document signing on the building's bell tower. The table to be used for this declaration of dependence signing would be from Abe Lincoln's law office with the desk from Thomas Jefferson's home. Lincoln symbolized national freedom from raw material needs, and Jefferson symbolized the link with agriculture. The inkwell was an exact copy of that used for the signing of the Declaration of Independence.

The paper was fine hemp paper, which had been used for the Declaration of Independence. Hemp was three times as strong as cotton, and Ford planted the hemp in nearby Greenfield Village fields for new industrial applications. Because Ford's village had a school system of over 500 kids, Ford's hemp fields required twenty-four-hour guard. Hemp was a variant of the illegal drug marijuana, and its production was highly restricted by the government. The scroll was titled "Declaration of Dependence upon the Soil and Right of Self-Maintenance." The declaration declared, "Through the timely unfolding of nature's laws, modern science has placed new tools in the hands of man which enable a variety of surplus products of the soil to be transformed through organic chemistry into raw materials usable in industry."[1] The American flag flying over the conference was made of hemp just like the earliest flags. Henry Ford always used visuals to make statements.

Not far was the building's 1929 cornerstone, which was a concrete slab with Edison's signature and the garden spade of Luther Burbank. Ford considered Edison and Burbank to be the nation's earliest chemurgists. Clearly, Ford was trying to weld the earliest crack in our nation's mission of Jefferson's agrarian view and Alexander Hamilton's industrial view for the nation. In many ways, Ford and the chemurgical movement looked to resurrect the old economic Whig Party of Henry Clay, which fused business and industry. The chemurgical movement was a strange mix of politics, industry, science, and agriculture. The movement strongly opposed the New Deal practice of paying farmers not to plant crops. Strangely, it also opposed the banks, and government alliance, which the movement saw as a major roadblock to both industry and agriculture. The opening speaker at the first Dearborn Conference declared, "American business has allowed the parasite of big banking to fasten itself upon it and to feed upon it and speak for it. This induced the government to turn parasite in turn, and American industry is caught between the two."[2] These extreme views and its politics would result in the movement's short life as a scientific movement.

The chemurgical movement was also an economic declaration of independence that the conference was to represent. Chemurgists maintained that the American farm was capable of making America free of dependence on foreign raw materials. Henry Ford,

1. Declaration of Dependence on the Soil

Harvey Firestone, and Thomas Edison had led a national movement to become independent of imported rubber. In 1935, America consumed 60 percent of the world's rubber and was totally dependent on imported raw rubber. During World War I, the nation had suffered from shortages as Germans cut world shipping and Mexican bandits closed down nearby alternatives. Between wars, the British cartel manipulated rubber prices. The rise of Nazi Germany had stirred new fears of being cut off from this strategic raw material. Edison had spent the last years of his life searching for an American substitute for rubber latex such as goldenrod. He had built a rubber laboratory at his Florida home with neighbors Ford and Firestone. In addition, Ford had invested in a Georgia plantation to experiment with alternative plants, and Firestone was looking around the world for new sources; but for Edison, it was a national quest for a North American source. Ford was now anxious to discuss with conference attendee George Washington Carver his belief that the sweet potato plant might be a source of rubber. Rubber would be a key topic on the 1935 chemurgical conference agenda. Carver had been working on sweet potato rubber since 1917 when Edison had wanted to discuss its possibilities.

The speakers and attendees, included many scientists and leaders such as Harper Sibley, president of the United States Chamber of Commerce; Roger Adams, president of the American Chemical Society; Frank Knox, publisher of the *Chicago Daily News*; Charles Stine, vice president of DuPont Chemical; Louis Taber, head of the National Grange; and W. Bell, president of American Cyanamid. Henry Ford had invited George Washington Carver to speak at this first conference, but he was unable to attend. The conference had department heads from universities such as Yale, Michigan, MIT, and Northwestern. The president of the Chemical Foundation and the Chemurgic Council kicked off the meeting with a chilling comment on America's dependence on Europe for rubber, chemicals, and dyes: "No matter who is elected President this fall ... that president will be compelled to dance to the tune of 'God Save the King,' and J. P. Morgan & Co., head and control of our New York banks and the head of the agents of foreign banking systems, will wield the baton."[3] These leaders feared dependency on foreign nations as the drums of war in Europe were being heard here.

These followers of the chemurgical movement opposed the New Deal programs of paying farmers not to produce crops; they wanted the surplus crops to be used for ethanol, industrial products, and chemical production for the nation. This unfortunate political angle caused the chemurgical movement to meet heavy resistance from the Roosevelt administration. Another concern of the Roosevelt administration about the chemurgical movement was its theme of nationalism about trade and the idea of supplying America's raw materials from internal sources. Most politicians of the time had turned to free trade as a necessity for the world economy and peace. Chemurgists favored protecting key American industrial and agricultural products and saw capitalism as a national, not an international, system.

The topics of this conference would be amazingly close to chemical, ecological, and green conferences of today. The keynote speaker of the conference was to be Dr. William Hale of Dow Chemical, who would introduce his future book of alcohol as the world's new fuel (*Prosperity Beckons: The Dawn of the Alcohol Era*). Ford was pushing an ethanol-gasoline mix to fuel his cars, but he was receiving heavy opposition from the oil companies. Some would present the results at this conference that a two-thirds gasoline, one-third ethanol mix was the "perfect" automobile fuel, having high octane ratings and antiknock properties. The conference even had a discussion of cancer being caused by exhaust fumes in burning gasoline. Ford had always believed the exhaust fumes were a major problem. As early as 1916, he had formed a company to produce clean vegetable-based fuels for his automobiles. Ford and the chemurgists had been gaining momentum until new finds of oil in Oklahoma and Texas drove the price of gasoline down to ten to fifteen cents a gallon. The best 1935 chemurgical processes for the manufacture of ethanol resulted in twenty-five cents a gallon.

Ford even took on big oil, saying that Standard Oil was blocking the development of ethanol. Ethanol was hailed as a clean-burning fuel, which had been the original fuel for the Model T. The conference would propose the ideal automotive fuel to be a 40 percent to 60 percent gasoline mix. Ford called this fuel Argol. Had the nation made that switch in 1935, history and the nation would be far different today! Ford was still talking of his old friend Thomas Edison's prediction that cars could be fueled by electrolysis of water as well. This conference was going to hear of the use of plywood and vegetable plastics in aircraft, new American sources for natural rubber, glass and plastics for safety windows, milk substitutes and plastics from milk, new fibers from vegetables for clothing, and breeding cotton plants for oil versus fiber. Many medical topics were also on the agenda.

Ford had lined up tours of his waterpowered village industries on the rivers of Southern Michigan to demonstrate green manufacturing. Conference attendees would see subassemblies and auto parts being manufactured by part-time farmers using green materials and clean energy. These small factories also processed soybeans from surrounding farms into plastic parts. These thirty small factories were Ford's vision of a farm-connected decentralized industry. Ford had truly developed a green supply chain and envisioned a farm-based car. Ford saw these village industries as a chemurgical solution to the nation's economic problems as well. In addition, Ford argued that these factories distributed more money to rural areas, increasing the standard of living.

Another tour for the attendees would be Ford's museum and Greenfield Village. Greenfield Village had working laboratories as well as a history of the manufacture of farm machinery. Ford would illustrate the long relationship of agriculture and industry with these reconstructed factories and farms. Furthermore, attendees would see Ford's improved system of education for future agricultural engineers and scientific farmers. Ford's museum and Greenfield Village were, in fact, a history of chemurgy and its future. Attendees were treated to chemurgical food and smacks such as soybean soup, ice cream, and cookies.

The conference ended with the formation of the National Council of American Agriculture, Industry, and Science. These industrialists and farmers were focused on the poor economic condition of the nation and saw chemurgy as an economic solution. The conference ended with a set of goals that were particularly geared for a nation in the Great Depression. The five points were to

1. Result in the gradual absorption of much of the domestic farm surplus by domestic industry.

2. Put idle acres to work profitably.

3. Increase the purchasing power of the American farmer on a stable and more permanent basis. And, thereby

4. Increase the demand for manufactured products, thus

5. Creating new work for idle hands to do; reviving American industry; restoring American labor to productive enterprise; and relieving the economic distress of the nation.[4]

There would be a 1937 conference that George Washington Carver would speak at that would again have over 300 attendees. Ford was happy to take guests to his new million-dollar soybean research in Dearborn, which he hoped would entice Carver to come to Dearborn. Meals at the conference included soybean soup, burgers, sauces, bread, cookies, ice cream, pie, milk, butter, coffee, and croquettes. This was similar to meals served at the Tuskegee Institute based on the peanut by Carver of peanut soup, peanut mock chicken, peanut coffee, milk, ice cream and cookies. Ford Motor Company handed out booklets on soybean recipes. Ford hoped to win over George Washington Carver from peanut and sweet potato products to his beloved soybean. Actually Carver had been a pioneer in soybean research but found the peanut and sweet potato more table ready for southern farmers. Ford was already funding research at Carver's old school of Iowa State. Ford Motor was running soybean-processing plants in Saline and Milan, Michigan. Ford even had a soybean plastic parts department at his Highland Park plant. Ford's research center had already created ink, glue, candy, plastic, paint, oils, fabrics, dyes, and industrial products. Ford scientists had created a textile fiber from soybeans. Ford showed up at one of the conferences with his clothes (except his shoes) made from soybeans.[5]

Ford had hoped the conference could interest the government again in rubber independence and the winds of war were in the air. Herbert Hoover, secretary of commerce in the early 1920s and later president of the United States, had been a strong advocate of rubber research and chemurgy. President Hoover had been the keynote speaker in the 1929 opening of Ford's Edison Institute and Greenfield Village. The takeover of the White House by Franklin D. Roosevelt had changed everything. Chemurgy turned political, as the Roosevelt administration called for crop reductions and less plant research. Farmers were to be paid for not producing versus finding new applications for the surplus. The idea of creating a farm surplus for new applications appeared threatening to FDR's New Deal farm policies. The first 1935 conference had created only fear in the Roosevelt administration of its political intents.

A follow-up pamphlet of the council in 1935 became more heated politically, stating, "So this is the vital issue before the farmers of America today: Shall they in the future operate their farms as they think best or shall they work under the direction of a federal bureaucracy with the power to regulate their lives, to levy overwhelming taxes and even confiscate their property." Also, the New Dealers had little time for the Republican business titans such as Ford. But the administration bias would even include the humble chemurgist George Washington Carver. Ford and the chemurgists had to work through intermediaries such as Eleanor Roosevelt and Secretary of Agriculture, Henry A. Wallace (and future vice president 1941–1945). Wallace, former student of Carver at Iowa State, had initially been ambivalent toward the movement, but he turned against fearing the Republican connections to the movement. Still, Henry A. Wallace had been a Republican and had strong ties with Carver. The chemurgists had also won over a young army major, Dwight D. Eisenhower, in 1930, who spent several months studying the strategic problems of the American rubber companies. In 1930, Eisenhower wrote a report advising the necessity of investing in chemurgical rubber research. Still, the politicians avoided any real commitment until Pearl Harbor.

Ford's travels in the 1920s and 1930s had brought together a core of chemurgists. Ford had spent the last twenty years forming his own Juno of creative minds, and this Juno would be incorporated in the chemurgical movement. His camping trips with Thomas Edison, John Burroughs, and Harvey Firestone had made headlines. The campers drew many presidents such as Calvin Coolidge, Warren Harding, and Herbert Hoover, which helped spread the movement in Washington. The group would often stop for visits at the homes of other inventors and creators such as Luther Burbank. Ford had continued a friendship and correspondence with Burbank, hoping to make him more active in his chemurgical movement. Burbank did get funding from Ford, but he never fully embraced Ford's industrial chemurgy. Burbank's interest was more in pure botany versus chemistry, and the chemurgical movement seemed to attract more chemists than botanists. Ford's chemurgical headlines were created by farming journals and rural reporters across the nation.

This 1937 chemurgical conference was to bring in more chemists than botanists. The term chemurgy was created from the Egyptian "chemi" and the Greek "ergon," meaning to put chemistry to work in industry for the farmer. The conference attracted many diverse groups, from big chemical scientists at Dow and DuPont to small farm groups. There was a group of interested politicians and even religious leaders attending this conference who had an interest in agricultural science. The farm lobby was well supported looking for help during the Great Depression. There was a segment of nationalists, "fair" traders, and protectionists seeing chemurgy as a means to national independence of foreign goods. Still, for many university-trained scientists, chemurgy was based in old methodology and lacked the status and scientific discipline of a true branch of science. And of course, there were the New Dealers who saw chemurgy as a threat to their price and production controls.

The 1937 conference would have Carver, and that would bring the national press; and Ford had some new ideas as well. He awaited conference attendees like an anxious father at his reconstructed Greenfield Village railroad station. He was going to unveil his new quest to make a completely green car at this conference. Ford already had a full-scale organic laboratory operating in Dearborn and manufacturing prototype green parts and products, mainly from soybeans. Ford had turned soybeans into paints, plastics, lubricating oils, soaps, enamels, varnish, ethanol, linoleum, and inks. Ford hoped to interest Carver in his soybean research, since Carver had done some of the original research on the uses

1. Declaration of Dependence on the Soil

of soybeans in 1904. There were to be many interesting new product discussions at the conference such as honey in golf ball cores, plastic food and drink containers, and what we would call today the engineering of plants for new uses. Ford, in particular, was looking forward to the talks of George Washington Carver on new uses of the sweet potato and peanuts. These applications ranged from industrial uses to a cure for acne. Ford also looked to discussing his soybean plastics with Irene du Pont of the giant chemical company.

Most attendees at these first chemurgic conferences were not young men; its two central attendees, Ford and Carver, were both in their seventies. Many in the scientific community looked at Ford and Carver as past their prime. Even Ford's son and his young Turks looked forward to the day of Henry's retirement to put an end to many of his projects. Most of these distinguished chemurgists had lived in and been part of the Industrial Revolution. The chemurgy movement was under attack from an unholy alliance of New Deal politicians, university scientists, atheists, the oil companies, bankers, free traders, the Roosevelt administration, large farmers, and the synthetic chemical industry. Both Ford and Carver were suffering from a decline in their scientific image from their association with the chemurgical movement; and in many ways, they represented the last of the Victorian scientists of the golden era with Edison and Burbank gone. These were men of the Victorian mold of Edison. Carver and Ford based their science on observation and experiments. They were hands-on, finding theory through experimentation. Like Edison, who hired a personal mathematician, they needed to look to others to help with the mathematics of modern science. They preferred their electricity from a chemical battery of the 1700s rather than from the new mechanical dynamos of electrical engineers. They preferred experience to the theory of the universities. Ford could not read very well the blueprints of the new breed of engineers, but created models to be his blueprint.

With his political, social, and cultural bias, Ford tended to rub university-based scientists the wrong way. Similarly, Carver's giving credit to God and calling the laboratory "God's little workshop" diminished his creditability with the new breed of university scientists. Even more disturbing to them was Carver mixing science and religion, which was reminiscent of Thomas Aquinas, Michelangelo, Newton, Priestly, and Albert the Great of the Middle Ages. Furthermore, there was a strong push back from the New Deal farm policy politicians and the oil companies who saw chemurgy as an opposing political and environmental movement. Ford appeared to be a traitor to the very industrial landscape and industry he had created. Some viewed Carver as a type of modern alchemist, since he often mixed prayer with his healing methods and spoke of finding methodology in the Bible. In reality, these were used as reasons to stop the chemurgical movement. Both men were true geniuses on the caliber of any university scientist. Regardless of Carver's inspiration and Ford's unusual views, few scientists and engineers have matched their results. Ford looked forward to having a friend to replace Edison in the movement.

These two leaders of the new chemurgical movement had admired one another but had never met. Henry Ford had admired George Washington Carver for years, and now Ford felt alone, having lost his buddy and muse — Thomas Edison. Carver had also been a friend and admirer of Thomas Edison. Edison had tried unsuccessfully to bring Carver to work with him at his Menlo Park research center. Carver had shared Ford's vision to bring industrial prosperity to the American farm. He had also been impressed with Ford's efforts to desegregate the automotive industry, allowing blacks into the crafts and apprentice programs. Certainly they had many shared beliefs, even though Ford had some problematic beliefs in several areas. Now in 1937, Carver and Ford would finally meet. Ford waited for the arrival of Carver at the Greenfield Village reconstruction of the Fort Smith railroad station that Edison had worked at as a boy. A few years after this meeting, Ford would similarly honor Carver by reconstructing his boyhood home at Greenfield Village. Ford, more than anyone, elevated Carver to the level of America's great inventors. This day he anxiously awaited America's best-known black scientist. Ford was also ready to make Carver a partner in his new movement.

While Carver was happy to travel to meet Ford, he came with many concerns about working with predominantly white scientists and businessmen. He had been invited to such scientific conferences in the south, only to find he could not eat at the banquet table. Some of these southern scientists would also be there in Dearborn. En route to a 1930 lecture in the west, Carver was refused to sleep in "white" cars. Often, the railroad might give in to a famous black man such as Carver but make it clear that other blacks must be segregated. Ford had sent a special railroad car for Carver and would tolerate not even the slightest hint of racism. Ford assured that Carver would stay at his five-star Dearborn Inn across from Greenfield Village. Carver was a humble man who often made his point quietly. At this 1937 Dearborn Conference, while Ford considered Carver an honored guest, Carver remained outside the main hall until all had eaten. Ford respected that decision and would assure all that Carver was the most important man present. It would be part of their future bond of friendship. Ford, for his part, had one of the best records on desegregation among America's industrialists.

As Ford waited at the Greenfield station, he looked forward to this great scientist who shared his vision. Edison's death had left Ford without a muse for his love of industrial chemistry. Henry Ford and Edison had been working at their Florida compound studying potential plants for rubber production. Ford had idolized Edison to the point of having Edison's last breath captured in a test tube for display at Greenfield Village. After Edison's death in 1931, Ford could hardly bring himself to travel to his Florida home where Edison was his neighbor. Ford saw in Carver a man who could replace his beloved friend. Carver's love of the peanut and its application matched Ford's own belief in the soybean as a universal material. Ford had also built a huge experimental farm, plantation, laboratory, and summer mansion for himself and Edison in Georgia, where he hoped Carver would join him.

For Carver, Ford would help him get the status of a great chemist who had eluded him. Even at his Tuskegee Institute, Carver was regarded as a "creative" or "cook stove" chemist to distinguish him from the theory-based chemists at the major universities of the 1930s. Ford, however, saw him as another Edison. Edison, too, had been a Victorian experimental-based chemist. The chemists of the 1930s were moving into the mathematics and theory of the science to plan their research. Edison was weak in mathematics and had hired a mathematician to do his calculations. Edison's methodology was compared

1. Declaration of Dependence on the Soil

to searching for a needle in a haystack. Yet scientists like Edison, Carver, and Burbank were masters of deduction, learning from failed experiments. Their success in these difficult searches was proof of their genius. Ford saw the genius in these Victorian chemists and scientists that was often lacking in the modern scientists of the day. The new breed of scientists needed less creativity and genius, using theory to help reduce the search. These new scientists loved mathematics, while men like Carver and Edison loved the feel of the physical world they studied. Where Thomas Jefferson once said he could not live without his books, Edison was likewise known to have said he could not live without his chemicals. Edison so loved his collection of thousands of labeled chemicals, he released them to Ford's museum only after his death. Carver, too, loved his collection of chemicals, minerals, and plant specimens.

Both Carver and Ford were experiencing a type of spring in their last years. Chemurgy was breathing new life into these old geniuses. This new fusion of agriculture and industry was seen as world changing. It was an area of science where old Victorian approaches could still be profitable, since organic chemistry was just beginning to be theory driven. For Carver, he envisioned that these new industrial uses of plants would bring wealth to poor southern farmers. Ford hoped to reverse the decline of rural communities in the north. Both men thought the combination of agriculture and industry would not only be a source of wealth for America but a source of food for the world. Both men saw it as necessary to a nation facing strategic shortages of rubber, oil, food, dyes, and plastic as the drums of war became louder. The dependency on rubber plantations in the Philippines had already created a panicked search for a replacement plant. Furthermore, the Germans held a monopoly on chemical dyes. Both Carver and Ford had been active in the First World War in helping America develop alternative materials and resources. Carver, in particular, made major contributions in breaking the German world monopoly on dyes.

Florida neighbors Ford, Edison, and Harvey Firestone had built a neighborhood lab in Fort Meyers to research new sources of plant latex to make rubber. Ford moved the Edison rubber laboratory to Georgia just before Edison's death. Ford had also developed his own rubber plantations in South America (Fordlandia) and was researching improved agricultural practices as well. Carver had been working on the sweet potato plant as a North American rubber source. Furthermore, Carver had been following the work of Edison using the Russian dandelion. In fact, prior to World War I, Edison had recruited Carver to join his rubber research. It was the type of challenge these men loved, but Edison's death had ended the quest. Carver was the perfect replacement (if not better) to lead this chemical quest. He had already developed a series of dyes, paints, and plastics from plants. Ford also envisioned, like Carver, that plants would lead a new industrial revolution in industrial materials. Both men wanted alcohol to replace gasoline as an automotive fuel. Much of the great work of these men is only now starting to be reviewed again. It was lost to the wonder of a new breed of scientists that created synthetic rubber.

The Ford and Carver union of 1937, however, expanded the overall chemurgic movement beyond pure industrial applications. Ford was interested in expanding the world's food supply, and Carver had spent a lifetime expanding agriculture in the south. Carver brought his friends such as John Kellogg into the movement. During World War II, Carver and Ford opened the National Laboratory for research in foodstuffs. They also teamed up at a Dearborn laboratory to continue Ford's search for a synthetic rubber, a project started years earlier with Thomas Edison. It was here for a brief period that the two experienced the fun of science known in their youth. Chemurgy offered new horizons for a country still struggling with the agrarian view of Jefferson and the industrial view of Hamilton. For a few brief years, Ford and Carver would once again be on the creative path. Their friendship was one of a shared love of industrial chemistry.

Both Edison and Ford had great respect for Carver. Ford saw Carver on the same level as Thomas Edison. Edison, of course, had long ago admired Carver, offering him $25,000[6] to come to New Jersey and work with him (he refused to stay in the south). After Edison's death, Ford lacked the synergy and symbiotic relationship that had inspired him. Eventually, the friendship forged at the 1937 conference would lead to a continuation of Edison's chemurgical projects. Ford built an experimental farm in Ways, Georgia, for Carver to work on. In addition, Ford built a trade school beside the farm for young blacks to learn a trade. He also established grants for the Tuskegee Institute, which still functions today. It was a friendship that both men had waited a lifetime for. And for a few short months in a Dearborn laboratory, these men in their seventies experienced the springtime of their youth.

2

Common Soil

"Birds are the best of companions. We need them for their beauty and their companionship, and also we need them for the strictly economic reason that they destroy harmful insects."

— Henry Ford

"I literally lived in the woods. I wanted to know every strange stone, flower, insect, bird, or beast."

— George Washington Carver

It might be easier to look to the differences of these two men, which were many, but that would not explain their friendship. Certainly, the differences were striking, with Carver starting life as a slave around the last year of slavery (1865) in the state of Missouri. Carver's birth date lacked the proper recording afforded by white Americans. As a baby, George and his parents were kidnapped; however, a Union scout was able only to bring the small baby back to the farm. Carver and his brother Jim would never know their real parents, only their adopted white parents, Moses and Susan Carver. Still, his adopted father, Moses Carver, was a fairly prosperous, pro–Union farmer who, prior to the war, had owned slaves to work his 240 acres. The farm grew both flax and hemp for fibers. After the war, Moses Carver would suffer greatly in the postwar economy. Still, Moses Carver's farm was almost self-contained. He grew food for the table; hay, wheat, rye, flax and hemp for clothes; and animals for meat, milk, transportation, and leather. The farm made its own soap, dyes, clothes, shoes, herbs for medicine, and its own animal feed. The farm had bees for honey production, and it managed a fruit orchard. Moses Carver was extremely proud of the self-sufficiency of his farm. The basic idea of the self-contained farm would remain with young George throughout his life.

Henry Ford, on the other hand, was born two years earlier in 1863 to somewhat prosperous farmers in Dearborn, Michigan, about ten miles from Detroit. The Ford family ninety-acre farm was smaller than that of Moses Carver; however, it was one of the largest in the Dearborn area. The Ford farm was typical of the Midwest, growing grain such as corn and wheat as well as raising dairy cows, poultry, hogs, and horses. Ford's farm was run for a profit versus for basic subsistence. Midwest farmers were self-reliant and made home products, but their farms neither were self-contained nor was this their goal.

Henry Ford, however, could claim humble roots. Henry's father, William Ford, had immigrated from Ireland in the great migration wave of 1847 driven by the potato famine. The only thing unusual was that the Fords were Anglican Protestants in the Irish county of Cork, which was 95 percent Catholic. The protestant roots came from the immigration of the Fords from England to be plantation managers of the Irish in the 1600s. Henry Ford would later favor a less rigid Christianity and be more nondenominational. It was far different from the deep Christian beliefs of George Washington Carver.

The difference might not seem so stark between Carver and Ford if you consider that George Washington Carver worked for open-minded and caring farmers in the state of Missouri, much like Ford's parents. Carver and Ford both had deep roots on the farm. The Carver family was childless and took on the responsibility to raise George and his brother. Young George found an interest in helping Susan Carver with her garden. She taught George how to cook, a skill which he would use throughout his life. Moses Carver farmed, raised livestock, and was a beekeeper and racehorse trainer. The family wove and dyed their clothing, made their own leather shoes, smoked meat, canned food, and prepared their own medicine. George Washington Carver would later reflect a similar diversity in skills. In addition, George found a love of painting, with a nearby neighbor who made her own paints from plants and soil. Painting would be a hobby he would cultivate over his whole life. More importantly, painting and the making of paints was the inspiration for the study of chemistry from his earliest days. The Carvers were not learned people but realized the importance of education for children. This importance would be a shared value of Carver and Henry Ford, even though their experiences were different. More importantly, both men shared a love of rural education.

On the Missouri frontier, early schooling was done at home using the old *Webster Elementary Speller*, which Carver would later say "he knew by heart," but he had no one to answer the book questions.[1] Common points for Carver and Ford would be the home Bible reading of their mothers and Sunday school. Like most farm-born Americans of the time, the Bible and *Webster's Elementary Speller* were the basis of initial schooling. Still, George had started at a white public school only to be rejected by other white neighbors. It was painful for George, but it did not diminish his love of learning. The experience would prepare George for a life of racial segregation. Moses and Susan, who lacked formal education, did obtain a private tutor for George and his brother. Black education had been mandated in Missouri, but each township had the option to segregate black students into black schools. George and his brother were getting an education from a circuit church with a mix of visiting Presbyterian, Baptist, and Methodist preachers. Early on, George demonstrated a high intelligence, and the Carvers looked for schooling opportunities for him.

George studied on his own. He often asked questions of any adult he could find. Church provided some reading skills in Bible study. Similarly, George developed at diverse set of skills ranging from knitting and cooking to beekeeping. He learned to play a number of musical instruments. Still, Carver seemed less interested in raising animals. It was clear that while George had an aptitude for gardening, like Henry Ford, he was not to see farming as his destiny.

When he was twelve years of age in 1877, the Carvers allowed George and his brother Jim to move eight miles to the county seat of Neosho. George would move into the black home of Mariah and Andrew Watkins in exchange for chores. In Neosho, his education would continue along many fronts. Mariah Watkins was an herbalist with knowledge of plants and flowers that had always fascinated George. In addition, Mariah taught George the art of cooking, a skill that would pay many dividends for him in finding work. His new black school, however, proved disappointing, with the teacher knowing barely more than George did. Within a year, George would be part of a greater migration of blacks to the state of Kansas. Kansas, the home of John Brown, was open to black Americans and offered a more accepting environment. However, as black immigrants poured into the state taking white jobs, the atmosphere turned bitter.

While Jim Carver was happy with the Neosho school in Missouri, George wanted far more. He moved to Kansas, again exchanging room and board for chores. Carver's cooking skills made him particularly attractive as a boarder to white families. With the family of Lucy Seymour, Carver picked up the business savvy of operating a laundry, as well as the basic chemistry of cleaning and dying clothes. He proved to be a young capitalist, shining shoes and doing other jobs to save money to open his own laundry business. While Carver's talents and work ethic opened many doors, he would also feel the pain of prejudice and the fear of black lynching in Kansas. In 1879, the young Carver would see a lynching firsthand that would haunt him throughout his life. It would be the first of many painful experiences with racism, but he never let it overcome him or allowed himself to be enslaved by anger; and in the end, he changed racial relations for the better.

The lynching motivated George to relocate to Minneapolis, Kansas, where he moved in with the black family of Ben Seymour. George found a variety of jobs to save money while attending school. In a predominantly white school in Minneapolis, George found acceptance and encouragement for his knowledge. By 1880 at age fifteen, he had saved enough money to get a bank loan to open his own laundry business in a small shack. He continued going to school and working for another four years. Carver proved frugal, finding ways to save on the tightest of budgets. He searched continuously for more and better educational opportunities. It would be in these Kansas schools that he would find another common ground with Henry Ford — the *McGuffey Reader*.

Henry Ford's early childhood showed a similar restlessness as George Washington Carver's, but a more favored path. Middle class farming in Dearborn was the stuff of Norman Rockwell covers. Autumn hayrides, winter skating parties, red barns, and iconic schoolhouses were the stuff of his childhood. His Scotch-Irish school was one of the best. Both Ford and Carver would have lifelong favorable memories of their childhood schools. Even though Carver would become a true scholar and Ford more the pragmatic engineer, they both had a deep love for this rural education. Where Ford and George Washington Carver would experience many similar roots was the Bible, *Webster's Elementary Speller*, rural one-room farm schools with good teachers, and *McGuffey Readers*. Both men would find lifelong enjoyment in these books. It says a lot about the McGuffey style of education that could bring pleasure to both the boy scholar and the typical playful boy.

For a young Ford, school was an escape from the farm chores. Ford loved the life of rural play and adventure. Like Carver, Ford had a fascination with nature and natural science. Ford cared little for the hard chores of farm life that Carver had no choice but to embrace. Ford hated the hard work of farming, and later in his 1922 autobiography,

The Green Vision of Henry Ford and George Washington Carver

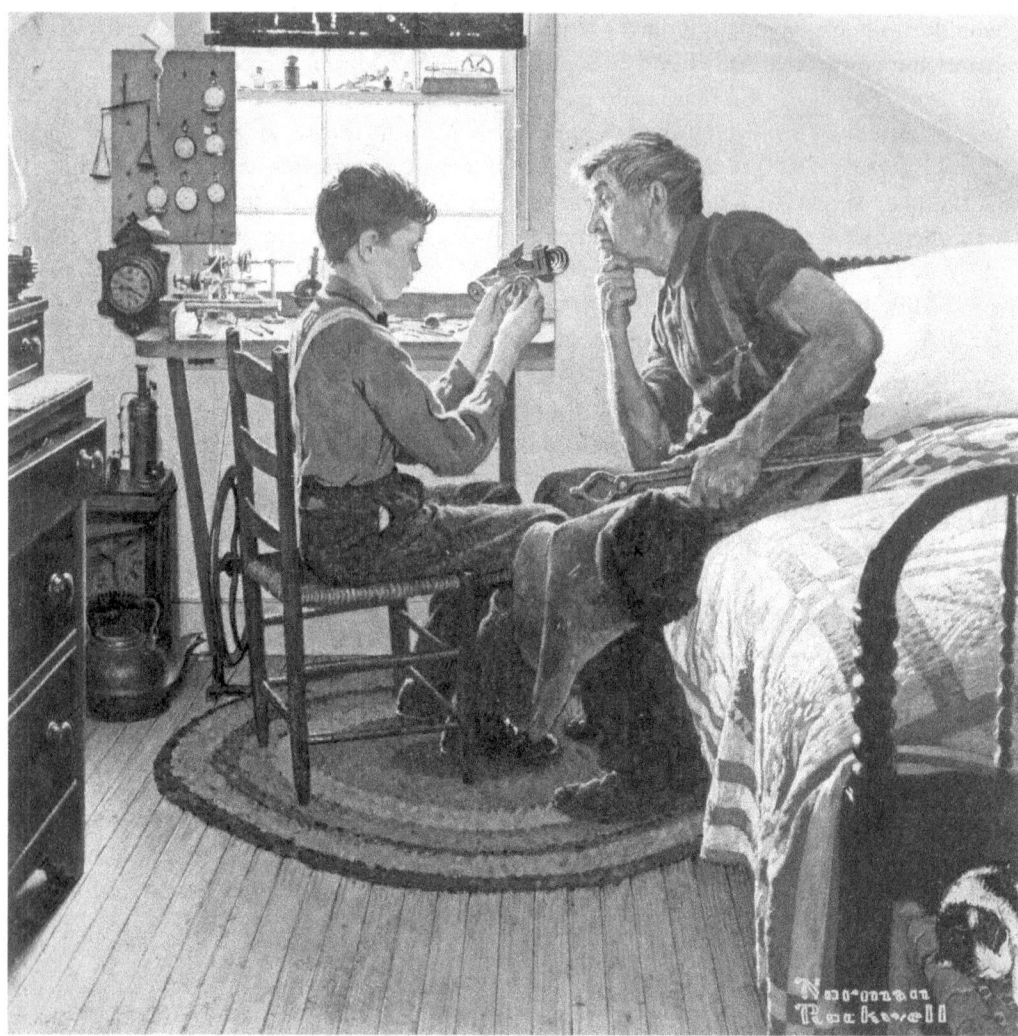

A Scene Depicting Henry Ford as a Boy Showing the Local Blacksmith a Model of a Wagon Rigged with a Clock Mechanism for a Power Unit. Oil, 16.9 × 16.4 in. A Norman Rockwell–commissioned drawing of a young Henry Ford at his workbench with his father (from the Collections of The Henry Ford, Benson Ford Research Center, Photograph THF95629).

he would state, "The farmer makes too complex an affair out of his daily work. I believe that the average farmer puts to a really useful purpose only about 5 percent of the energy that he spends."[2] Ford would be nostalgic about the farm but never cared for the work. He did find a love in the farm mechanics of the equipment. This was the passion that Henry took to as well as the popular science of the times. Ford and his friends built a small grinding mill using a waterwheel and a steam engine from their reading of various boy-oriented books and magazines. As a middle-class son, Ford was able to move on from the farm with the blessing of his father. For Ford, his early farm days (less the chores) were one of his beloved memories. Ford would become passionate about giving as many boys as possible the rural education of his youth.

America's greatest teacher was William McGuffey (1800–1875), whose famous *McGuffey Readers* dominated America's one-room schoolhouses for most of the nineteenth century. Henry Ford championed William McGuffey's teaching approach and his *McGuffey Readers* until his death. In a 1936 *New York Times* interview, Ford lamented, "Observing the type of character produced in the schools of the McGuffey period, I am convinced that we must seek in our educational methods the causes of some of the faults today.... Today there are too many frills in education.... When this country adopted a system which put less stress on moral principles, the children grew up and seemed like ships without rudders."[3] Industrialists and leaders educated by *McGuffey Readers* make up a long list. These Gilded Age leaders often had nothing more than the four readers in the McGuffey series as their education. The most interesting factor is the shared vision of these McGuffey-"trained" industrialists, blending agriculture, nature, and industry. George Carver was also a strong supporter of the McGuffey methodology.

Interestingly, both Carver and Ford believed in the strict McGuffey classroom approach as well. It was an approach that demanded memorization, direct questioning, and discipline. Even Ford, who often wore the "duns cap" and sat in the corner with his friend Edsel Ruddiman, would later see the need for discipline. Later in life, Carver would disagree with his Tuskegee Institute boss over the need for McGuffeyism. He would write the following in 1902: "Students want to discuss the topic rather than give a direct answer. This is not permissible. Nothing is more to be deplored in the classroom to hear pupils giving their opinions."[4] Carver would be a strong supporter of McGuffey's approach of including nature study at the early level. Still both men believed in a hands-on approach to learning as well as training in the trades.

In many ways, Ford remained a third grader his whole life, often breaking words into syllables and collecting sayings in endless little notebooks. Henry Ford credited McGuffey for his success and turned part of his personal outdoor museum of Greenfield Village into McGuffeyland. Ford so loved his *McGuffey Readers* that he became an avid collector of the many editions. Today, Ford's museum has over 300 editions (the largest in the world). In the 1930s, Ford started the quest to bring McGuffey's home to Greenfield Village. In the search for McGuffey's home, he found a debate in Washington County as to whether the Blaney or the Lockhart farm was the birthplace of McGuffey.[5] Like St. Helena with the early Christian sites in fourth-century Jerusalem, Ford deemed the Blaney farm the birthplace by edict. Henry Ford moved the log cabin of the Blaney farm, brick by brick and log by log, to Greenfield Village in 1934. In addition, he shipped old oak logs on the site to build a model McGuffey schoolhouse beside the home at Greenfield Village. This "McGuffey" schoolhouse included a teacher's desk used by McGuffey. Ford turned all his Greenfield Village sites into active schools for Detroit children.

While in the home township of McGuffey (Finley, Washington County, Pennsylvania), he got involved in an effort to save one of Pennsylvania's earliest covered bridges. He ended up bringing that bridge to Greenfield Village as well. To further increase McGuffeyland, Ford brought his Scotch Settlement School in Detroit, where he was introduced to *McGuffey Readers*, to Greenfield Village. He added ancillary buildings such as the home of his favorite schoolteacher. Greenfield Village remains today a tribute to the role of William Holmes McGuffey in educating the American industrialist. In the 1930s, Ford started his own school based on the *McGuffey Readers*, using many of the buildings in Greenfield Village as classrooms. In addition, Ford would bring Noah Webster, the author of *Webster's Elementary Speller* (and dictionary) from New Haven, Connecticut, to Greenfield Village as well. Ford even went as far as to bring a young nephew of McGuffey to Dearborn to attend his school.

Both Ford and George Washington Carver showed the high morals and the desire to achieve associated with students of the *McGuffey Readers*. The McGuffey approach was rooted in the concept of self-reliance by Ralph Waldo Emerson. *McGuffey Readers* can be credited with much of the success of the Gilded Age. Abraham Lincoln called McGuffey "Schoolmaster of the Nation." President Harry Truman praised the *McGuffey Readers* for "educating for ethics as well as intellect, building character along with vocabulary." Presidents see better than most the problems of society, and they realize the limitations of government in trying to correct them. The readers often quoted and used sections from Emerson's "Self-Reliance." McGuffey was certainly consistent with Emerson's notion of self-reliance, but he was also consistent with Emerson's belief that education must precede politics. For both Emerson and McGuffey, politics dealt with the symptoms of societal problems while education dealt with the root causes. Ford often carried quotes from Emerson's "Self-Reliance" throughout his life. The success of early American business was based on a moralistic compass that allowed for limited government intervention. McGuffey's moralistic approach went to the heart of culture by preparing a moral citizen. McGuffey prepared a generation of politicians and businessmen that at least knew what high ideals were.

While Ford hated farmwork, he was entrenched in the lore of American farm life often found in the *McGuffey Reader*. He was a walking farmer's almanac of weather and crop sayings, superstitions, and prejudices. He worried all his life about walking under ladders or in front of black cats. When he saw a red-haired man, he looked for a white horse. He was said to hate to leave the house on Friday the thirteenth. Some of the old farm cures and food might explain his strange interest in diet and food. He loved to read the essays of Emerson such as "Self-Reliance," which he was first introduced to in the *McGuffey Reader*. A more recent biographer, Greg Grandin, stated, "Much of Ford's faith that industry and agriculture could be balanced and that community would be fulfilled rather than overrun by capitalist expansion drew specifically from Ralph Waldo Emerson."[6]

While only an average student, Ford had fond memories of his Scotch settlement grade school and teacher, John Chapman. And his desk mate, Edsel Ruddiman, would be a lifelong friend. Edsel and Ford loved to play jokes and get into trouble. Years later, Ford had the Scotch settlement school torn down and reconstructed in his outdoor museum of Greenfield Village. On the reopening of the old school in Greenfield Village, Ford and his friend would once again take their seats for reporters on hand. Ford not only collected over 300 editions of the *McGuffey Readers* but read them often throughout his life. Ford would also build his own private school system for poor kids based on the McGuffey system.

One continuing myth is that Ford picked up his anti–Semitic views from the *McGuffey Reader*.[7] McGuffey stressed Christian virtues over dogma but used direct biblical quotes that some consider put Jews in a bad light. Some biographers have attributed the anti–Semitic views of Henry Ford as being rooted in his McGuffey experiences, but there are no concrete examples to make this linkage. The fact is that McGuffey led the movement for better treatment of Jews, Catholics, and Indians, which had all been targets of nineteenth-century discrimination.[8] The *McGuffey Readers* stressed diversity and tolerance. If McGuffey's readers showed any bias, it was in the very early editions with slavery. Henry Ford, of course, deplored slavery and discrimination against blacks. In reality, the southwestern Michigan society of Henry Ford's time, while Protestant and abolitionist, had bias against Jews and Catholics. The dislike of Jews was related to the common misconception that they controlled the banks, which were the farmers' archenemy. Any linkage of the McGuffey readers to any type of bias has little credibility.

Both Carver and Ford developed a love for nature through the farm and the stories of the *McGuffey Reader*. The *McGuffey Reader* had numerous readings on nature and animals, which were of interest to the frontier child of the west. The readers injected the mystery of exotic animals such as elephants, giraffes, and lions. This mix of nature helped draw in young boys to read and study. Nature study was the basic science of the day and would develop a general interest in science in Ford and Carver. Their interest in nature would be an integral part of their lives and careers. Most McGuffey-type schools of both Ford and Carver focused on nature subjects like tree, bird, and insect identification. Later in life, George Washington Carver would publish a booklet for schoolteachers in 1910 titled "Nature Study and Gardening for Rural Schools," and a 1902 booklet titled "Suggestions for Progressive and Correlative Nature Study." These were very similar to the early writings of William McGuffey seventy-five years earlier.

As a young child, Carver loved plants and flowers. He collected leaves, rocks, insects, and bird eggs. He was always asking about their names and uses. Carver also loved to collect rocks, insects, and tree leaves. Carver was known to have large specimen collections as a youth. It was a hobby he continued throughout his life. Carver loved vegetable and herb gardening from an early age as well. Similarly, Ford exhibited a passion for identifying birds and was able to identify many by their calls. Ford started as a youth to record the return dates of various species and continued to keep records until his death. In later life, Ford was known to have over 2,000 birdhouses on his estates. He used heated birdbaths in the winter to attract birds. He often passed out signed birding handbooks to children and friends. He also helped pass the Weeks-McLean bird bill in 1913 providing bird sanctuaries throughout the nation. He would be elected president of the Michigan Audubon Society and personally supported birding programs for children and an annual bird census. He even imported hundreds of birds for release on his farms.

Unlike Carver, Ford's education ended at grade school as he moved on to follow his passion in mechanics. While Ford's father hoped that he would follow him as a farmer, he understood and shared his son's love for things mechanical. The steam engine tractors of the nearby farms fascinated Ford. Machinery was the one item that Ford believed should have been incorporated in schooling. Later in his own school system, he added the study of mechanics to the McGuffey curriculum. He loved machinery of all types. He particularly became adept at fixing mechanical toys and watches. Ford became popular in the neighborhood for fixing timepieces. He even made his own tools such as tweezers and screwdrivers from pieces of scrap iron, corset pins, and nails. As soon as his father would let him, Ford started as a machinist apprentice at Flower Brothers' machine shop in Detroit. While renting a small room, Ford only stayed eight months there. So while Carver was moving from state to state looking for better schools, Ford was moving from company to company in search of more money and opportunity.

Ford next moved to the machine shop of the Detroit Dry Dock Company. Detroit Dry Dock was one of the largest shipbuilding companies on the Great Lakes. It was hard and long work. Ford learned not only the basics of machining, but the operation of steam power used to drive the machines. A typical shift was ten hours, with pay being two dollars a week. Like Carver, Ford's various skills made him very employable. In addition, he worked nights at Robert Magill's watch repair shop for fifty cents a night. While he loved to repair watches, it was clearly a necessity since his room and board was $3.50 a week.[9] Henry Ford's passion for mechanics grew, as he loved reading magazines such as *Scientific American*, *World of Science*, *The American Machinist*, and other trade magazines for machinists. Ford set up a small workshop behind his boardinghouse to run a lathe and tinker. He remained restless, dreaming of inventions and of making more money. Ford worked long hours, but unlike Carver, he had time to enjoy his weekends with his school and lifelong friend, Edsel Ruddiman, and his new fellow apprentice, friend Frederick Strauss.

3

Finding Their Mission

"And God said, Behold, I have given you every herb bearing seed, which is upon the face of all the earth, and every tree, in which is the fruit of a tree yielding seed; to you it shall be for meat."
— Gn 1:29, Carver's life verse often quoted

Ford and Carver had both shown an unusual work ethic coupled with ambition to succeed. Their youth consisted of long hours and hard work. Still both found time for their hobbies and passions. Both were looking for a career that could weld their passion and their need to be successful. Both retained their youthful dreams throughout their lives. As both Ford and Carver approached their late teens, their future paths became clearer but not straighter. Both remained restless, often taking a rogue path versus a straightforward career path. Both found encouragement to follow their dreams. Like all boys, they had dreams, but theirs differed from most in that they would not compromise their dreams. Both now would take youthful detours and find many forks in the roads of passion and necessity. This was what made them restless, unwilling to make a youthful necessity in life a lifelong career. Ford and Carver would change jobs at an extreme rate even for today, let alone for the Victorian age, where career selection or apprenticeship was made very early in life. As a biographer, it is hard not to see the hand of Providence.

There is no question George had the harder path with the racial prejudice of the times. Yet Carver's is a story of overcoming each racial roadblock and one of inspiration. Carver even used the rejection as fuel to move on and achieve. While he found acceptance at his white school in Minneapolis, Kansas, he found racial prejudice in the society in general. He was refused service at local restaurants and had to ride in separate railroad cars. Even in his very successful laundry business, he had to assure white customers that he washed only the clothes of whites. Signs had to be posted: "We Wash for White People Only." Still, the struggles seemed to strengthen him. He became more determined to be successful. George expanded his skills into learning and playing many musical instruments, sewing, cooking, crocheting, and becoming active in the Presbyterian church. He added typing to his long list of skills so he could find more work. In a way, George was progressing quickly toward some yet unknown destiny.

Years 1883 through 1885 would be ones of pain and rejection for George Washington Carver. He returned to Diamond Grove to visit the Carvers, only to find that his brother Jim had died of smallpox. He had been close, and Jim was his only known blood family member. George recalled many years later how his faith carried him through this loss. He moved to Kansas City and found a job as a clerk because of the typing skills he had learned at the Union Depot. He now felt truly alone, but he kept moving forward. The real personal blow came in 1885 when he applied to Highland College by mail. His writing skills showed, and he was accepted. It was a proud moment for a twenty-year-old young man who had dreamed of being a scholar. He used all his money to move up to the northeast corner of Kansas. When he presented himself to the college's president, he was refused entrance based on his color. It had to be a crushing blow to a young man's dream. It clearly was the type of blow that starts one questioning if the path is right. Many that have experienced such dream-ending realities become depressed or resentful. Carver had to give up his education for a time and looked to a new direction in farming and homesteading; but it failed in the end to fully extinguish his passion for scholarship.

George did try a new path after his rejection at Highland College, having no money to even leave town. He had no choice but to find work near Highland College. He worked briefly for the white family of George Steeley, gardening and cooking. Carver once again was able to integrate into the area's white society. He took painting lessons, played the accordion for dances, joined the local literary society, and built a garden of note in the town. Townspeople were amazed at Carver's knowledge of plants and nature. He was not only making money again but saving it. At least he seemed comfortable. Carver, however, remained restless. He was searching for an opportunity that could satisfy his creativity.

The Homestead Act had opened up a much different opportunity for Carver. In the fall of 1886, Carver was able to put a down payment on a 160-acre section of land on the nearby western Kansas frontier. Carver's frugal saving had been enough for the down payment, but he had little else. His personal belongings were an accordion and a watch. The prairie lacked timber and bricks, forcing Carver to build a sod house. He planted ten acres of corn and a vegetable garden so he could be self-sufficient. The second year he planted fruit trees such as plum, apricot, and mulberry. He drew heavily on the thrifty and practical farming taught to him by Moses Carver. Carver even returned to his painting. He didn't know then that this would be an important learning experience for his later career. Carver was one of the rare black settlers, but he clearly won the respect of his white neighbors. So impressed was the community that a story was published in the *Ness County News* in March 1888.[1] The article noted that his knowledge of "geology, botany, and kindred sciences is remarkable, and mark him as a man of more than ordinary ability." Carver had been one of a handful to survive the winter of 1888, but prairie farming was not to be his destiny. His hobbies remained more reflective of his future.

Carver not only collected rocks and minerals but clays and soil, which he used to produce dyes and paints. He found a special interest in clays because he produced an array of paint colors from them. Clays gave him natural colors of yellow, red, and brown. He got blue from processing plants and made green from mixing yellow clay with this bluing. This bright blue pigment would later attract scientists, industrialists, and an Egyptologist, who thought he might have unlocked the secret of the rich blue of ancient Egypt. One variation of purple Carver himself called the "lost purple of Egypt." Carver's deep blue dye and paint would be hailed by many and copied by many as well. These earth-based colors were reminiscent of the impressionist painter who similarly loved the rich

3. Finding Their Mission

clay-based colors. Carver's earth- and clay-based colors were noted around the world as he continued their development throughout his life. He probably did as much with the uses of clay as with the peanut. Painting would always lead Carver back to science. He developed a method to make his canvas from cornstalks. Carver became a prairie scientist. The article further noted his geological collection and "conservatory" of over 500 plants. Lacking books, he often developed his own classification systems for plants and rocks. Even on the prairie, he was a Victorian renaissance man. The article implied that if not for his color, he would "occupy a different sphere."

Carver managed to find kindred spirits on the prairie. One of these was Clara C. Duncan, an art teacher from Talladega College, who helped Carver improve his drawing skills. George particularly took an interest in the drawing and painting of flowers and plants. He would excel in the painting of flowers over his life. Clara also helped George with poetry and English composition, igniting his fire for education and further study. The Great Blizzard of 1888 convinced him that homesteading was not his path, and at the age of twenty-three, he once again pulled up stakes and moved to Winterset, Iowa, to become a food buyer for a hotel and open a laundry. Winterset would later become famous as the birthplace of movie star John Wayne.

Carver's friendly personality again allowed him to assimilate quickly into the local community. Carver's church attendance would bring him in contact with the area's most prominent citizens. His musical and artistic skills impressed all. Now Carver would get the encouragement he needed to return to his goal of attending college from friends at his church. He was also impressing many with his painting and drawing skills. He gathered the courage to resume his education at the age of twenty-six by entering Simpson College at Indianola, Iowa. It would take courage as Simpson College's first ever and only black student. Carver would once again demonstrate his ability to break down racial barriers and stereotypes. Furthermore, his classmates were sixteen and seventeen years old, much younger than Carver. Simpson College was a liberal arts college, lacking the science courses he loved. The money was also a struggle, requiring long hours of laundering. However, this time he found an open and helpful organization at Simpson. Students helped him with housing, furniture, and money to buy books. In the faculty, he found encouragement for his art. His studies included art, piano, essay writing, grammar, and arithmetic. Carver excelled at painting and art, which were his majors. His art teacher, Etta Budd, was impressed with what she called his natural ability in painting especially. College was a struggle for him while working. At college, however, was finally where he belonged. But he was not at the right physical institution.

George was again encouraged to focus on his passion in agriculture by becoming a scientist. Etta Budd, his friend, was a daughter of a professor, J. L. Budd, of horticulture at Iowa State University, who realized that Carver had the aptitude for botany and more scholarly research. In 1891, Carver once again packed up and moved to Iowa State College of Agriculture and Mechanic Arts. The college would soon become known as the agricultural Harvard of the Midwest. Iowa State was a first-class agricultural school with three of the nation's four future secretaries of agriculture from 1906 to 1932 graduating from the school.[2] Providence would bring all of these important figures into Carver's life.

Future secretary of agriculture James Wilson arrived at the Iowa State Ames campus the same year as Carver to run the government experimental station. Carver would once again be the first black on campus, and he met with resistance from some students. James Wilson, however, was quick to bring Carver to the experimental station. Wilson had been elected as a Republican to the House of Representatives and was a close friend of a future president, William McKinley.

Initially, George was not allowed to sleep in the dormitory or eat in the dining hall at Iowa State. He persisted with the help of old friends to open doors. His personality soon won him many friends, and friends from Simpson College supplied money to help out. Iowa State would prove to be the cornerstone of Carver's career. He established himself as a scholar, and then Providence brought him into contact with a number of future leaders of agriculture. Men like James Wilson honed him in the principles of "scientific agriculture."

Carver participated in a diverse array of activities including German Club, YMCA, Art Club, Welsh Eclectic Society, school plays, the Agricultural Society, football team trainer, and the Student National Guard. Carver found happiness in the National Guard, loving the uniforms. He also enjoyed the punctuality, routine, precision, and neatness. He rose through the ranks to captain. Carver enjoyed chemistry and geology courses as well. He excelled at botany, publishing an article titled "Grafting the Cacti" in the *Transactions* of the Iowa Horticulture Society in 1893. He followed with an article a year later titled "Best Bulbs for the Amateur." His senior thesis, "Plants as Modified by Man," would augur his future as a chemurgist and agricultural expert. His thesis was an extension of the work of Luther Burbank. The thesis was a study of cross-breeding and grafting of commercial trees and flowers. Carver had become an expert in the improvement of plants through hybridization. He continued his paintings, winning local and state acclaim. At a state exhibit at Cedar Rapids, students pitched in to send him to see his *Yucca and Cactus* take state honors. The painting was one of several to represent the state at the Chicago World's Fair.

It would be at the World's Fair in Chicago that George Washington Carver and Henry Ford's path would first cross. Ford proved just as restless during this period. He was a successful urban apprentice in 1882, but torn between the country and city. Even the Detroit area itself seemed to struggle between the two. He clearly wanted both, so when the opportunity came to return to his father's nearby farm, Ford jumped on it. Ford's father had hoped to entice Henry back to the farm. Ford's return to the farm was to help a neighbor with his new Westinghouse steam-powered grain thresher. Ford made a hefty three dollars a day working with the thresher during the summer.[3] He soon was the area's expert on heavy farm equipment.

For a short period while living on his father's farm, Henry Ford worked for the Westinghouse Engine Company as a traveling troubleshooter. His father added some land to interest Ford in farming and helped him set up his own machine shop in the barn. A lot of mechanics, a laboratory, and a little farming minus the chores were the perfect combination for Henry Ford. He pursued his interest in birding, taking long walks and looking through his binoculars. The shop consisted of a blacksmith's forge, a drill press, a lathe, and masses of hand tools. As a hobby, he repaired watches. He then bird-watched in the meadows. It was a type of Eden for the young Ford. He started his own experiments to build a steam engine tractor, using parts discarded by surrounding farmers. His efforts and repair work soon attracted other farmers to use Ford's services. He even attracted the regional manager, Charles Cheeney of Westinghouse Company, which supplied steam equipment to farmers. From 1883 to 1885, Ford traveled Michigan repairing and demonstrating steam-powered equipment.

These were times of great advances personally for Ford. They mixed machinery, study, and a little farming. He studied competing equipment of the German company of Daimler, expanding his knowledge of gasoline engines. He spent his spare time expanding his home shop and working on his own steam-powered equipment. In the winter months, he attended Goldsmith, Bryant, & Stratton Business University. Here he took courses in business and mechanical drawing, which he disliked. The bookkeeping courses were the root of his passionate cost cutting throughout his career. He enjoyed socializing and would meet a hometown girl, Clara Bryant, and the relationship would result in an 1888 marriage. His father gave him an eighty-acre farm to settle, hoping to entice him to become a farmer. He used his new mechanical skills to build a steam-powered sawmill, which became the nucleus of a striving business. The lumber business was booming, with area shipyards, factories, and farms expanding. Ford worked summers, servicing engines for Buckeye Harvester.

Meeting Clara slowed Henry down a bit. He was actually becoming quite the dancer and social party boy. He was well known for his appearances at the Greenfield Dancing Club. Henry and Clara enjoyed picnics, dancing, and long walks in the woods. They both loved music. They joined the local chapter of the New York Chautauqua movement known as the Bayview Reading Club. This adult education movement would have a major impact on Ford. The marriage was a simple one in the home of Clara's parents in Detroit (Greenfield Township). William Ford was hopeful that the now-married Henry would settle into a life of farming. Henry, however, was still not fully convinced that farming was his future.

Ford enjoyed the mechanical challenges of the repair business the most. He was an avid reader of *Scientific American* and other engineering journals, and he dreamed often of inventing. These magazines and journals were filled with the work and person of the electrical wizard and inventor Thomas Edison. Edison became an iconic hero to Ford. Ford dreamed of heroic breakthroughs in mechanics. Henry Ford was rapidly becoming the regions expert in engines and fuels. Few people at the time had his experience in various types of engines. He first worked on an Otto gasoline engine in 1885 for a farmer. This one-cylinder, four-cycle combustion engine offered advantages over steam. The Otto engine required an electrical system. Ford lacked any knowledge of this, but this lack of knowledge would lead him in a new direction.

Electricity was the high technology of the day. Ford's lack of knowledge and restlessness had him once again looking for a job in Detroit. Ford wanted to find a way to learn the new technology. This time he applied for and secured a machinist job at the Woodward Avenue substation of Detroit Edison Illuminating Company. In 1890, the plant supplied most of Detroit's residential electricity. It would be the ideal mix of machining and learning electricity as well as applying his knowledge of steam engines, which were used to generate electricity. It was an opportunity to work for a personal hero, Thomas Edison. The job also paid well at $480 a year. The Detroit Edison plant would be a critical step in the career of Henry Ford, and years later, he would have the plant fully reassembled in his outdoor museum of Greenfield Village.

In September of 1891, Henry and Clara packed up once again to move to Detroit. They rented a house for $120 a year. His supervisor was his old boss at Flower Brothers, and his salary quickly increased to $1,000 a year as he became chief engineer at Detroit Edison. That was a large salary for the times, but the work was hard. Ford worked the night shift and was on call. It was critical in this new electrical power generation industry to keep things running, and Ford had already played the role of troubleshooter, getting farm engines running to make tight time windows for planting and harvesting. The heart of the Edison operation was its huge 100-horsepower Beck steam engine. Ford quickly learned the electromechanical operation of the dynamo and generator.

The rapid promotion was a result of Ford's experience in steam power. The plant had been built in 1886 using a large steam engine to generate electricity. Ford proved his skills with a number of steam engine failures, requiring timely repair. He became known in Detroit industry for his repair and troubleshooting, which brought in more money. Other manufacturing companies that were having major problems often borrowed him. Still, Ford brought his workshop and hobbies to his Detroit home on Michigan Avenue. He seemed sure that even his successful career at Edison Illuminating was not his final destination. Henry's success allowed him to work a less demanding schedule, playing the role of troubleshooter versus supervisor. Managers seemed to accept Henry's slack time because of the need for emergency maintenance. For this emerging industry, emergency repair was a necessity for success. Ford also became more valuable for his knowledge than for his hands. For a restless and wandering youth, such working conditions were ideal. Ford had the time and equipment to pursue his tinkering and inventing.

As his interest in a gasoline car increased, Henry found more "slack time" to work at his home workshop. Ford's youthful friend, Fredrick Strauss, described his routine: "Henry never used his hands, to tell the truth. He never came to work until after nine o'clock either. He was a free lance at Edison."[4] In 1891, he became interested in bicycles, whose popularity was sweeping the nation. The time was not all for tinkering; like Carver, Ford filled his days pursuing his interests. Ford taught machining classes at night at the YMCA for $2.50 a session.[5] The YMCA offered access to the newest equipment for Henry as well. He considered briefly motorizing the bicycle but remained focused on the horseless carriage.

It is doubtful that Henry Ford was fully focusing in 1892 on a horseless carriage as some more hagiographic biographies suggest. Like all great corporate owners later in life, Henry Ford had the ability to embellish events of the past. It is clear he was dreaming of being an inventor in the general field, as were many young men of the time. Stories of his hero Thomas Edison's life from rags to riches were everywhere. Ford did enjoy long bicycle rides around Detroit and worked on bicycle improvements. Henry was also con-

3. Finding Their Mission

tinuing his work on the gasoline engine and had built a small inefficient one in his workshop. The quest for a gasoline engine in general had been the topic of many scientific magazines of the time, which were Ford's favorite reading. The year 1893 would bring the goal of a horseless carriage more directly into Ford's life. At the Columbian Exposition in Chicago, Ford would see his future.

4

In Search of Destiny at the Fair

"Pray for me please that everything said and done will be to His glory. I am not interested in science or anything else that leaves God out of it."
— George Washington Carver

"Everything is in flux, and meant to be. Life flows. We may live at the same number of street, but it is never the same man that lives there."
— Henry Ford

The Chicago World's Fair (known as the Columbian Exposition) of 1893 would be a key moment in the lives of both Ford and Carver, although they never physically crossed paths there. In fact, Carver spent little time at the fair, but his painting, *Yucca and Cactus*, did win awards. For Carver, it was a badge of courage in overcoming racism and a lack of money. For Henry Ford, the Chicago World's Fair was the equivalent of an eight-year-old going to Disneyland. Both men drew inspiration from this great event as did much of America. The greatest of the Victorian scientists, engineers, and businessmen were there. Men like George Westinghouse, Thomas Edison, Edward Libbey, H. J. Heinz, Nicola Tesla, Henry Clay Frick, Michael Owens, B. F. Goodrich, and Gottlieb Daimler were there. More importantly, the Chicago World's Fair offered a uniquely American vision fusing agriculture and industry. This was much different than the industry and science focus of some earlier fairs. There would even be new products introduced using peanuts—Cracker Jack and Kellogg Peanut Butter.

The Chicago World's Fair was to be called the Columbian Exposition in honor of the 400th anniversary of Christopher Columbus's landing in America in 1492 (to be held a year late due to the presidential election). The Chicago site covered more than 700 acres and was enough for over 60,000 exhibits. One out of every four persons in the United States traveled to Chicago to see the fair. A total of 27 million visitors from all over the world passed through the gates. Its cost of $25 million would still bring in $2.25 million. The famous Ferris wheel, designed by George Ferris, could carry 2,000 people at once. There were thirty-six trolley-size cars with forty stools (and like in a trolley, some passengers stood). The ride cost fifty cents, and with loading and unloading, the two revolutions took twenty minutes. The Ferris wheel dominated because of its 250-foot steel structure. It was the favorite ride of most of the fairgoers, and it was estimated that 1.5 million people rode it. There were over 65,000 exhibitors, and many would make their name at the fair such as Juicy Fruit gum, Kodak box cameras, typewriters, gasoline engines, Heinz ketchup, Heinz baked beans, Aunt Jemima syrup, Libbey Glass, Gillette's safety razor, Jell-O, Possum coffee alternative, Cream of Wheat, Kellogg's Shredded Wheat,

4. In Search of Destiny at the Fair

George Washington Carver in later years showing his 1892 World's Fair entry *Yucca and Cactus* (Bentley Historical Library, University of Michigan, File HS8575).

and Pabst Beer. The fair introduced diet carbonated soda, hamburgers, exotic dancing, and postcards.

The fair would herald the birth of electrical power, featuring arc lighting, incandescent lights, electric trains, electric boats, electric motors, electric cranes, lighted fountains, telephone service, electric fire alarms, and endless futuristic appliances. This was the greatest fair of mechanical Victorian devices, making it a pilgrimage for Henry Ford. It would herald the end of the Edison direct current power station of the type Henry Ford had been chief engineer at in Detroit. The Chicago Fair was the major battle between the Edison DC system and the Westinghouse AC system. The inherent advantage of AC current was the ability to carry it long distances versus having an Edison DC plant every few blocks. Edison had 121 central power (DC) plants throughout the United States, but primarily in densely populated areas that lowered its distribution costs. In about a year of market operation, Westinghouse had 68 AC central operating plants, plus many under construction. Engineers like Ford came from the world over to see the AC system of the Chicago World's Fair.

The Westinghouse powerhouse at the fair was revolutionary, and the lighting plant was the largest power station in the world in 1893. This idea of practical alternating current was Westinghouse's innovation, and it was part of his determination to fully develop and apply an alternating current induction motor. It was the engineering triumph that Tesla had theorized and Westinghouse Electric brought to reality. Thomas Edison, who used direct current to power his electrical devices, had fought alternating current at every step. For Ford, Edison was an idol.

The alternating current allowed for high-voltage transmission, which allowed for few high-cost power stations. Westinghouse transformers stepped down the voltage at points of application. The new rotary converters were used to change the alternating current (AC) to direct current (DC) for use by the electric railway, which required DC current. A number of the Tesla alternating current motors that were being used and were exhibited would be the basis of Westinghouse's future electric motor business. While Edison would lose the "battle of currents" to Westinghouse, the Edison lightbulb could run on AC current and would dominate the market. Ford would spend hours studying the generation of alternating current at the fair.

It was not the revolutionary generation of alternating current that occupied most of Ford's time though, but the exhibits in the Transportation Building at the fair. While locomotives dominated, there were several horseless carriages. There were three manufactured by Gottlieb Daimler as well as the Benz carriage that Ford had already seen. There were also electric vehicles that resembled motorized bicycles. The exhibit proved inspirational to not only Ford, but also to fellow inventors Ransom Olds and Frank Duryea. Ford would return to Detroit with a clear goal to develop a commercial horseless carriage.

For George Washington Carver, the Chicago World's Fair proved just as inspirational. One source reported that George carried the painting to the World's Fair himself,[1] while others even suggest he did not make the trip. The evidence is clear that Carver attended the fair as part of the governor of Iowa's color guard.[2] Certainly, the award for the painting boosted his morale after years of overcoming prejudice; but it would be his degree from Iowa State that would be his real accomplishment for the period. In addition, Carver would become friends with James Wilson who worked at the Iowa Experimental Station. He and James shared a deep commitment to Christianity and mission work. They also shared a love of music. Besides science, he and Wilson worked together at the YMCA to help new students. He often talked on a variety of subjects including race relations, which would become a theme later in life for him. As a devout Christian, Carver considered a career as a missionary. Later in life, Carver found his missionary work in helping blacks in the South. His proudest moment came in 1894 as he received a bachelor of agriculture degree. Still, art was in his blood, and his success at the World's Fair almost created another diversion. He pondered an offer to attend the Chicago Institute of Art. By now, Carver was not only accepted at the university but considered a major asset, and the faculty persuaded him to stay on as a graduate student.

James G. Wilson from Iowa State served as secretary of agriculture under William McKinley, Teddy Roosevelt, and William Taft (Iowa State University Library, Special Collections, Box 21, Photograph 1578).

Several very important figures were at Iowa State at the time. James G. Wilson, future secretary of agriculture under presidents William McKinley, Teddy Roosevelt, and William Taft, who was in charge of the Iowa State Agricultural Station, and had brought Carver in as an assistant. Another important professor was William Cantwell Wallace (1866–1924), who would serve as secretary of agriculture under presidents Warren Harding and Calvin Coolidge. Professor William Cantwell Wallace was a major influence on Carver with his mix of science, religion, and agriculture. Wallace's father was a well-known minister who also wrote farm pamphlets. William C. Wallace brought spirituality into the laboratory, which Carver fully adopted. Wallace also was responsible for converting Carver to "scientific agriculture" and the use of fertilizer. William C. Wallace and Carver would remain friends throughout life, corresponding until their deaths.

In addition, William C. Wallace would often bring his young son, William A. Wallace (1888–1965), future secretary of agriculture under Franklin D. Roosevelt and vice president and secretary of commerce under President Harry Truman, to work. Carver would teach young William A. Wallace, and they remained friendly but never truly close. However, William A. Wallace would break with Carver over New Deal politics in the 1930s. In general, Iowa State had a strong mix of religion and science, which became Carver's hallmark. Carver would always have strong ties to Iowa State.

The university made Carver an assistant in botany and manager of the greenhouse, which finally freed him from years of menial labor to pay for his education. At the Iowa Experimental Station of the government, Carver would publish a series of bulletins to help state farmers. He continued to publish for the Horticultural Society articles such as "Best Ferns for the North and Northwest." His graduate research was under Louis H. Pammel, an authority on mycology (plant diseases and fungi) and president of the Iowa Academy of Science. Mycology and hybridization would be of interest to Carver throughout his career. He quickly established himself in the field of mycology with papers titled "Treatments of Currants and Cherries to Prevent Spot Diseases," "Fungus Diseases of Plants of Iowa," and "Inoculation Experiments with Gymnosporangium." Dr. Pammel would always consider Carver one of his most brilliant students, and they would exchange letters and specimens. Doctor Pammel would remain Carver's mentor throughout his life.

Carver was a skilled plant botanist at this point in his career, and had he stayed at Iowa State, he surely would have been a university scientist of note. Carver was rapidly becoming an important mycologist and was being referenced in several scientific journals. Still, Carver was a Victorian scientist pushing the scientific advance with collecting and classifying versus theory. Carver would base his master's thesis on Luther Burbank's hybrid technology. Throughout his life, Carver would share with Henry Ford his admiration of Luther Burbank. Carver also did extensive work with Iowa State professor J. L. Budd, chair of Horticulture. Carver honed his skills under Professor Budd in hybridizing fruits. Professor James Wilson, future secretary of agriculture, said, "In cross-fertilization ... and the propagation of plants, Carver is by all means the ablest student we have here."[3] In addition to research, Carver taught classes and held Bible studies.

When Carver received his master of agriculture in 1896, he was once again confronted with a number of career choices. Iowa State clearly wanted him to stay. Louis Pammel was particularly interested in keeping Carver at Iowa State; however, Carver had several offers to move on. A black college in Mississippi, Alcorn Agricultural and Mechanical College, offered him a position with a salary increase. Carver represented the very rare black that held a higher degree in agriculture, making him a popular candidate for black colleges. Furthermore, Carver had considerable following and respect with white scientists. Had Carver stayed at these northern universities, there is no doubt he would have been a great scientist and probably an expert in the field of mycology. However, it would have been unlikely he would have dined with presidents and great industrialists. More importantly, he would have never impacted this nation at the level he did. Selecting a career in science at Iowa State would have been the safest path, and at the time the smartest; but Carver proved, as before, that he was a risk taker in search of a destiny. Eventually, he was drawn to a black college in Alabama, Tuskegee Institute (known as the Tuskegee Normal and Industrial Institute). Tuskegee was located on an old cotton plantation and hoped to offer training to the children of former slaves in scientific farming. On one level, Carver's appointment to Tuskegee Institute was an end of an educational quest, but it

4. In Search of Destiny at the Fair

would soon prove the next step also. At the same time, Henry Ford was also taking another big step.

Ford had started his quest for a gasoline-powered car in earnest on returning from the World's Fair, advancing to a climax in 1896. Ford began to spend more and more time at the workshop behind the Edison Illuminating Company with a group of young mechanics with a similar passion. These included competitor Charles B. King, who would be the first to put a car on the streets of Detroit; Ed "Spider" Huff, future racer and electrical ignition expert; Oliver Barthel, a machinist trained by Ford at YMCA; boyhood friend

An Interior Scene of the Bagley Avenue Shop Depicting Mr. Henry Ford Working on His 1896 Model Car while Mrs. Henry Ford Sits By. Watercolor, 16.3 × 16.6 in. A Norman Rockwell–commissioned drawing of Henry Ford, Clara Ford, and his Quadricycle (from the Collections of The Henry Ford, Benson Ford Research Center, Photograph THF95604).

Fredrick Strauss; and Edison Illuminating employees, George Cato and Jim Bishop. The group would meet on Saturday nights to test engines as well as gathering nightly. All described these meetings as fun because of their passion. Ford's farming background had given him knowledge of gasoline engines. Gasoline engines were first used on the American farm for pumps and for the grinding and threshing of grain. The only thing Ford hated about the gasoline was the noxious fumes. With Ford's knowledge, the group took on the challenge of building their own gasoline engine from the blueprints supplied in an 1895 issue of *American Machinist*. It would be Charles King on March 6, 1896, who would first take to the streets with the engine on a 1,300-pound wagon that barely reached five miles per hour, with Henry Ford following and observing it on his bicycle.[4]

Ford would test run his gasoline car, the Quadricycle, on June 4, 1896. In August, Ford would meet his idol, Thomas Edison, at a company convention in New York. Edison would hear of Ford's gas-powered car over dinner, but it was Ford who would never forget meeting Edison. Ford remembered Edison to say, "You have it." For Ford, it was a turning point he would later immortalize in a painting by Irving Bacon, which had been seen in Greenfield Village today. Years later, Edison and Ford would become best of friends. The Quadricycle used gasoline, but Ford could make adjustments to have it run off kerosene or even grain alcohol, which were readily available at the time. Gasoline was a nuisance by-product of kerosene refiners which fueled the lamps of America. It was the lighter distillate with explosive-type burning that made it a great combustion engine fuel, but eliminated it from use in oil lamps or as heating oil. Henry Ford's Model T would make gasoline the prime product of the oil companies. Ford would recommend that grain alcohol be mixed with gasoline to eliminate engine knock. Throughout his life, Ford would advocate alcohol and gasoline mixes as a way to improve the lot of American farmers.

Edison was the Victorian-type engineer and scientist that Ford (and George Washington Carver) looked to as a standard in methodology and as an iconic hero personally. He would touch the lives of both Ford and Carver in a very direct way. A prolific experimenter, Edison was practical but not interested in theoretical science. Edison was a hands-on inventor, building theory through endless experiments. Edison collected and loved chemicals as Carver loved and collected rocks, plants, and soil. Ford collected tools and metal. Edison had a love for chemicals and their magic that remains unmatched. In the late 1920s as Henry Ford rebuilt the extensive laboratories of Edison in Greenfield Village, the last thing Edison gave to the exhibition was thousands of bottles of his cherished chemicals. Edison loved the visuals and smells of chemicals. His direct current electricity was chemical driven from batteries. Edison lacked the math skills to fully understand the nature of the mechanical-driven electricity of Westinghouse and Tesla. Edison, Ford, and Carver were all Victorian, hands-on inventors. The laboratory and the workshop, not the theory, were their proving grounds. Science without the laboratory was of little interest to these Victorians. These three men would represent the peak of the Victorian scientist as well as the last of them.

5

Tuskegee and Detroit

"When you can do the common things in uncommon way, you will command the attention of the world."
— George Washington Carver

The year 1896 had been momentous for both George Washington Carver and Henry Ford. At least geographically, Carver and Ford had ended their roaming. Their careers were also finally set. Both men had overcome the obstacles that had haunted them for years. They were beginning to move onto the national scene. Greats like Edison and Booker T. Washington were starting to take notice. Still, their greatest accomplishments were ahead of them. The next ten years would be ones of taking care of business, but other passions would still stir in both of them.

His decision to go to Tuskegee was not an easy one for Carver, but it would define him. He had two good offers at well-known white schools. Tuskegee Normal and Industrial Institute was an all-black school that had only opened in 1881. "Normal" meant it was a teacher's college. The students had built the school using bricks, boards, and nails all manufactured by the students. Carver had a missionary zeal for the improvement of blacks based on personal experience. His prized possession was the bill of sale of his mother to Moses Carver. Still, Carver looked at Tuskegee as a short-term mission, not the start of a lifelong career. He hated the injustice of slavery as well as the economic injustice of his age, and Tuskegee offered a way for him to help change things. Tuskegee, however, came with many professional challenges. Carver would have to start the agricultural college from scratch and build the buildings using students. There was no laboratory equipment. Carver would lack the great laboratories of Iowa State. His friends and associates gave Carver a microscope as a leaving gift; it would be the first microscope at Tuskegee.

Carver had been offered a salary of $1,000 a year, as good as the other offers, but he was heading for a school that lacked even the most basic equipment. The $1,000 salary was much higher than the national average of $470 a year and a great deal higher than his associates received at Tuskegee. For Carver, it was a decision to help poor blacks gain an education; he had struggled against much prejudice to obtain his degree and now he wanted others to have an education. Carver was the only black scientist in the nation with advanced degrees from white schools. In his acceptance letter, Carver noted, "It has always been the one ideal of my life to be of the greatest good to the greatest number of my people possible and to this end I have been preparing myself for these many years; feeling as I do that this line of education is the key to unlock the golden door of freedom to our people."[1] Carver had never been an angry man, but he wanted to take that hurt and turn it into something positive. He would overcome prejudice by advancing the black man. What he was not ready for was the professional jealousy he would be confronted with from his fellow professors at Tuskegee.

Tuskegee offered every opportunity for helping the black man that other offers did not. Tuskegee was small, but a school rapidly gaining a national reputation under its founder and president, Booker T. Washington. Tuskegee Institute had been the vision of Booker T. Washington who wanted newly freed slaves to obtain the skills to live as free men. It had already gained a national reputation in the building trades. Tuskegee was in the middle of the Cotton Belt where most blacks were poor sharecroppers. Booker T.'s vision was to develop a school of agriculture and an experimental farm station to help blacks become prosperous farmers. He hoped that George Washington Carver would supply the leadership for both.

Booker T. Washington had started Tuskegee in 1881 around an old church in rural Alabama. Booker T. had come from a mining family to enter the Hampton Normal and Agricultural Institute in Virginia. The Hampton Institute and Booker's experience there became the model for his Tuskegee Institute. He brought with him the same emphasis on personal cleanliness, virtue, discipline, punctuality, courtesy, and practical skills. Hampton Institute had been part of the opening of education to blacks by federal legislation. Civil War general Samuel Armstrong was the president of Hampton, who opened it to blacks. His military background added a routine of military discipline and character. Hampton, and later Tuskegee, was a strange mix of discipline, including a 5:45 A.M. personal hygiene inspection for students. When the state looked to expand its normal school system, Armstrong recommended Booker T. Washington, who was an outstanding student leader at Hampton. Booker T. Washington came to Tuskegee in 1881. Tuskegee was to be an all-black normal school (teacher's college), but it was merely an old cotton plantation in 1881. Washington was forced to build the institute with students, although he did get federal and state aid.

Tuskegee's real growth and national fame came in 1895 with a controversial speech known as the "Atlanta Compromise." Booker T. Washington, playing to an all-white state exposition audience, announced that blacks should not look to politics or government but succeed through hard work. The speech did not go down well with many who had risked their lives to fight for the rights of blacks. Still, the speech did bring in large donations from philanthropists. Washington received a large influx of money from the Samuel Slater Fund. Samuel Slater had been a famous New England industrialist in the textile industry. Andrew Carnegie and John D. Rockefeller followed with yearly endowments. When Carver arrived at Tuskegee, he would stay in the newly built Rockefeller Hall Dormitory and hold Bible studies in the new Carnegie Library. Carnegie gave $20,000 to build a library in 1902. Both Rockefeller and Carnegie had committed to $10,000 a year, and a few years later, Andrew Carnegie would give $600,000.[2] Even Ford was giving to Tuskegee as early as 1906. There were smaller gifts of barrels of clothes and books coming almost weekly from northern groups.

Tuskegee was an amazing school that proved self-sufficient and profitable in many endeavors. There was no school equal to it in the trades. It was a lean operation with self-

5. Tuskegee and Detroit

Carver (front row center) and Tuskegee Institute faculty, circa 1902 (Library of Congress).

manufacturing, recycling, and saving. The school's brick-making operations, foundry, lumber mill, blacksmith shop, buggy shop, print shop, tin shop, bakery, and farm all supplied products for building and repair as well as for sale in the community. Students were paid for their work. The income allowed students to work to pay for their education. In addition, students produced their own clothes, did the laundry, and farmed and prepared meals. The students ran their own bank. The students made their own mattresses, pillows, brooms, tablecloths, sheets, and baskets. This became a trade school known as the Mattress Factory. Like Ford's early factories, Tuskegee was an organization with a mission. Teachers of even purely academic subjects were expected to visit working shops weekly to tie practical items into their lessons. Teachers often wore many hats including working in the trades. This type of self-sufficiency and education impressed Henry Ford a decade later. While Carver did not fully subscribe to Booker T. Washington's beliefs, he did believe that there had to be economic help to maintain educational equality in black colleges. Outside income built libraries, laboratories, and classrooms.

Carver would face more personal challenges, however, in moving to Tuskegee in the "Black Belt" of Alabama. The Black Belt was part of the Deep South where former slaves were now improvised tenant farmers. Macon County, Alabama, where Tuskegee was located, was at the heart of the Black Belt. Tuskegee was Old South in the heart of slave country. Carver would find a deeper and angrier prejudice than he had in the Midwest. In addition, the soil was poor and had been overplanted for cotton. Cotton had kept blacks in a form of economic slavery. The poor soil offered small yields, which were too small to pay the rent to white landowners. Another group was sharecroppers who farmed the land for a share of the crop. Often a sharecropper might get as little as four cents a pound for cotton while having to pay fifteen cents a pound for meat at the store. The landowners wanted cotton planted, so Carver had to change the minds of the landowners and sharecroppers. Carver had noted early in 1899 that "the average southern farm has little more than about thirty-seven percent of a cotton crop selling at four and half cents a pound and costing five to six to produce."[3] Thus, the black farmer was trapped in a cycle of poverty. Carver saw his contribution to his fellow blacks as the leader to free them of their economic chains. It was the most challenging problem of the post–Civil War era.

White landowners were at first very skeptical of Carver because his approach might take away from cotton planting. The landowners needed the economic returns of cotton to allow the black farmers to pay rent. The end of slavery had put strains on their costs as well. This is why in crops like peanuts Carver had to sell the economic returns for the white landlords. Carver had to educate them that the morale of the black farmer was key to their success. It was the same in Ford's car factories; morale was the real key to productivity. Black farmers who could barely feed their family could not be fully productive. Economic slavery was not an efficient operation, but this would require some real convincing.

The problems were not all external either. Carver had been used to white schools that, in the case of Iowa State, had been very supportive of his personal success. The faculty at Tuskegee was skeptical of Carver because of the hype and his salary. His salary of $1,000 was far above the school average of $400, which did not endear him to his associates. For perspective, consider that the average United States skilled worker made about $390 a year and a southern farm family of five earned $310 a year. In addition, Carver had won the position over a temporary incumbent, George R. Bridgeforth, which would be a source of endless friction. Carver had also made it known that he saw the assignment as a type of temporary mission to help and then move on. Booker and Carver both had egos that would often clash as well. And then there was the school culture of routine, discipline, and rules, which required endless administrative reporting. On top of all this, Carver lacked equipment and funding. The experimental station he was to manage was underfunded at a mere $1,500 a year.

Carver did like the military discipline and found comfort in his surroundings. Tuskegee, like Hampton, was built on military routine around a series of seventeen bells. Students arose at 5 A.M. at the rising bell and retired at 9:30 P.M. with the closing bell. Students had to clean their rooms for inspection before going to breakfast. They served on guard duty at night. Students wore military-like uniforms (made in the student tailor shop) and performed military drills. Nightly chapel was part of the daily routine. Food production and preparation were in-house duties required by students. Carver found out

5. Tuskegee and Detroit

on arrival that the management of this would be part of his duties. As a bachelor teacher, Carver was required to live with a roommate in the dormitory. Eventually (over three years) when Andrew Carnegie donated a new library to Tuskegee Institute, Carver was given two rooms in the old library. He set up a small art studio there as well. He proposed and was allowed to teach a painting class for students. He taught them to make their own canvas from plant stocks and to grind clay into oil paints. He dressed in old suits with ties made from cornstalks and dyed with his test colors. He became known for the flower in his lapel.

Most associates saw Carver's room more as a Carnegie museum. Carver brought his huge collection of fungi to Tuskegee. He continued to grow his soil, clay, and rock collections, adding a meteor he found near Tuskegee. He had plant fossils, dinosaur bones, and mastodon teeth. He practiced taxidermy with his students, having stuffed ducks, birds, wild animals, and even barnyard fowl. Wall cases were a mix of books and specimens. He had a leaf collection as well as tree limbs in every available corner. He collected all types of moss. He had glass specimen cases for his favorites. His office fulfilled the characterization of the Victorian scientist painted by authors such as Jules Verne, H. G. Wells, and Arthur Conan Doyle. He was part Professor Challenger of *The Lost World* and part Professor Liedenbrock of *Journey to the Center of the Earth*. Carver had the same disregard for his clothes as these Victorian professor stereotypes as well.

Carver's salary, dress, and office would become secondary to his own perceived attitude. The first few months would test Carver's resolve as seen in the following part of a letter to the school's finance committee: "I do not expect to teach many years, but will quit as soon as I can trust my work to others, and engage in my brush work, which will be of great honor to our people.... At present I have no room even to unpack my goods. I beg you to give me these, and suitable ones also.... At present the room is full of mice and they are into my boxes doing much damage I fear. While I am with you please fix me so I maybe of as much service to you as possible.... I am handicapped in my work."[4] Such complaints would hardly endear him to his associates. It was far from a happy beginning to a lifetime at Tuskegee.

However, Carver's presence at Tuskegee paid early dividends for Booker T. Washington and the institution. When Carver finally got the agricultural building complete, he asked his former Iowa State professor James Wilson, now secretary of agriculture for the McKinley administration, to attend. Wilson had been one of Carver's favorite teachers at Iowa State. Wilson and Carver had also been active in Bible study at Iowa State. In December of 1897, Wilson, President William McKinley, and the entire presidential cabinet, which included Teddy Roosevelt, came to Tuskegee to dedicate the Slater-Armstrong Building. It was the first time a major federal representative had visited Tuskegee Institute. Local white town officials joined Carver and Booker Washington in greeting the train. It was the biggest event the small town had ever known. It was a great celebration with a victory arch and fireworks. This was the type of publicity that Washington lived for. Carver also was a very popular teacher, which again gave him leverage in these endless administrative feuds. Still, it did little to smooth the relationship between Carver and Washington.

The struggle with Booker T. Washington would be the theme of Carver's life until Washington's death in 1915. There was a mismatch of priorities, expectations, and skills between the two. Carver had hoped the planned agriculture station at Tuskegee would be a center for research, as had been his experience at Iowa State. Booker T. Washington saw little need for pure research, wanting to focus on practical education. Washington expected Carver to fully implement and build the station and agricultural department, but Carver lacked the administrative and organizational skills to meet this expectation. Carver had a hatred for what are today common administrative duties at colleges, such as committee meetings, reports, written assessment, and bureaucratic paperwork. It was bad enough that the differences between the two men created problems, but the similarities caused even greater problems. Both were extremely sensitive to criticism and both found pleasure in exploiting their own strength to target others. Both men had strong vision and mission. They were ambitious and very proud of their accomplishments. Both men had large egos and a need for praise of their work. They became great because their visions were larger than the feuds, setbacks, and roadblocks.

Biographer Linda McMurray characterized the clash as follows: "Washington ran his school like the 'master of the plantation,' but Carver refused to play the role of humble slave."[5] Washington demanded detailed operating reports on the farm as seen in the following letter: "I want the milk report every morning, and on it I want the number of cows milked.... I must insist that it be sent promptly each morning and marked as I directed."[6] Other memos included observations by Washington of equipment left out in the rain, working students too many hours, pictures not properly hung, and faulty door fasteners. This back and forth makes one wonder how the two remained on campus, but Carver was too big a name to push too far. Washington understood there were limits to how much he could demand of Carver. Interestingly, Carver shared the same hatred of farm "chores" that Henry Ford did. A Carver biographer described it this way: "Part of Carver's ongoing problem with Booker T. Washington and the administrative process at Tuskegee was Carver was transparently delighted about subjects or tasks he enjoyed, and equally transparent in resisting work he disliked."[7] Much the same was said of Henry Ford in his early days.

Carver and Washington were in total agreement with the mission of Tuskegee in improving the lot of black farmers, and it was this that bound them together. Booker T. Washington had been working with eastern philanthropists to finance ownership of small black farms. The Southern Improvement Company was the financial arm of the effort. This organization formed in 1901 was highly successful in increasing black farm ownership and getting good mortgages. Still, Booker T. Washington needed Carver to change the agricultural methods to make these farms profitable. The two men found ways to work together by understanding their strengths and weaknesses. It was clear that they needed each other to achieve their goals.

Carver even started Bible studies for students using his free time, which Booker T. Washington had suggested. This was incorporated with the formal course of Bible study for training missionaries. Carver became active in the student chapter of the YMCA. Students readily accepted him, where the faculty stood back from embracing him. It was only a short time before he was winning over the hearts of the farmers as well. Carver, however, struggled with administrative duties and was slow to adapt to the infighting of academic campus life. Carver had been restless throughout his life, and it was difficult to make a long-term commitment. When faced with things he disliked, he had always moved on. As a graduate student in Iowa, he had been shielded from the politics of university life, and he struggled with his Tuskegee associates. Praise from peers comes harder than as a student.

When at about the same time Carver got an offer from Knoxville College to teach agriculture and art, James Wilson suggested he remain at Tuskegee. Wilson had political reasons to keep Carver at Tuskegee. The McKinley administration needed Carver in the Black Belt of the Deep South to change the economic dependence on cotton. For all the fighting and clashes, both Carver and Washington needed each other to achieve their goals. In many ways they were stuck with each other as well. Washington could bring in the money to finance Carver's science and help give Carver creditability in the scientific world. For Booker T. Washington, he needed Carver to lead an educational revolution in southern farming. The first few years appeared to be a contest of how far they could push each other. Washington made moves to ensure that Carver realized who was in charge. Carver believed God called him to Tuskegee, so he accepted his cross in the long run. Washington came to realize that Carver was not interested in taking his administrative job as president. Carver came to believe that his calling was bigger than the disagreements with Washington. After many fights, they found ways to share the glory and success at Tuskegee.

The relationship between Carver and Booker T. Washington did grow over time. They came to understand that they were not competitors, but they complemented each other. Booker T. learned that he functioned well with the social "400" of New York and dining with Teddy Roosevelt at the White House. While both men believed in the mainstream approach to race relations, Booker T. was the politician; Carver wanted to prove the point through science. Both men had felt the effects of slavery and discrimination, but Carver was never comfortable leading any social or political movement. As Booker T. Washington approached the twilight of his years, he and Carver became known for taking long walks when Booker T. had trouble sleeping. When Booker T. died in 1915, ex-president Teddy Roosevelt came to Tuskegee for the funeral, and it was Roosevelt who would pass the torch of race relations leadership to a reluctant Carver. By the time of Washington's death in 1915, Carver and Washington had become friends and found ways to live with each other. Neither made their differences public, but friends outside the campus were aware of the tension.

Amazingly with all the infighting, Carver demonstrated some very positive results at his experimental station: "He took a plot of land that was 19 acres of the worst land in Alabama to experiment on to find what could be done to improve production. The first year it brought him a net loss of $16.25 an acre. After his year of scientific treatment and cultivation it showed a profit of $4.00. With another year the profit was $40.00 an acre and every following year brought better returns." This was the type of research Carver lived for.[8] In the 1890s, Booker T. Washington had hoped that he could improve the life of the black farmer by increasing productivity. Carver's experimental station was an old cotton plantation, and the soil had been depleted. When he arrived in 1896, Carver focused on the development of fertilizers and crop rotation. He studied nitrogen-enhancing plants such as crimson clover, cowpeas, soybeans, peanuts, and velvet beans. With clover, Carver argued that beekeeping would help add additional income. Carver's father had been a beekeeper, and he increased honey production at Tuskegee with huge savings. Honey was a great sweetener substitute for sugar. Sugar had always been a major cost to the simple farmer and his family. Carver started a highly successful beekeeping operation at the Institute but found resistance in converting cotton farmers to beekeeping.

Soon Carver realized that clover was far from the answer to soil improvement, as it offered less opportunity as a food and cash crop. Carver also realized that the long-run decline in cotton prices and the slowly advancing boll weevil made the problem one of more than poor soil. The answer was in the planting of legumes like beans and peas. With most legumes, Carver needed to find new uses so they could be more economical in crop rotation. With the exception of the peanut, most people in the area had never seen these plants before. Carver started to show farmers the results from simple crop rotation plans. In 1901, Carver started extensive experiments in the nitrogen-producing crops of cowpeas, sweet potatoes, and peanuts. He preferred these nitrogen-producing crops to alfalfa and clover of northern farmers because these three produced an additional bounty of useful food. Carver also argued for a two-crop-per-year rotation where possible — fall plowing of summer crops of corn, cotton, peanuts, etc., and spring plowing in winter crops such as clover, wheat, soybeans, and cowpeas.

There was another success that both Carver and Washington could agree on. Carver started a series of agricultural bulletins that became very popular with Alabama farmers. The first bulletin in 1898 was *Feeding Acorns*, which advised farmers to mix acorns with expensive corn feed to save money. Acorn starch had been used by Indians for centuries as a food stock. Furthermore, Carver's experiments harkened back to his father's farm where pecans, walnuts, and hazelnuts were often fed to animals. Carver reported his results at the experimental station in feeding 400 head of hogs. Carver's research showed better milk from cows and egg production from chickens from the cheaper mixture of acorn feed. Acorns were a half feed because it had about half the food value of corn for an equal quantity, but acorns were available at a fraction of the cost. This type of new thinking and education was what Booker T. Washington and the secretary of agriculture wanted for the old Cotton Belt. Carver proved to be an outstanding science writer, combining popular science with the hard science of chemical analysis.

His second research-based bulletin was *Experiments with Sweet Potatoes*, which suggested a fertilizer mix that improved both yield and overall cost of production. Sweet potatoes offered a food source for the family as well. In 1898, Carver's experiments showed an amazing yield of over 250 bushels per acre. While overlooked at the time, sweet potatoes had a long history. Dried sweet potatoes were a staple for starving soldiers during the Civil War. Carver was also aware that Indians had made dyes from them. Another bulletin, *How to Build Up Worn Out Soil*, became the basis for saving many of the state's farmers. Carver detailed that plowing should be seven to nine inches deep with soil building materials such leaves, corn stocks, moss, and other plant matter being worked in. Carver complained at the ignorance of local farmers in building the soil back after usage. He further argued for resting the soil by quoting the Bible's Mosaic Law to rest the land every seven years. He also studied plants like okra, which had been grown by slaves for food. Carver used the okra fiber to create wear-resistant rugs. Only recently have engineers found that the tensile strength of okra fiber is extraordinary and started using it in composite materials.

But even Carver's bulletins came under attack by his colleagues at Tuskegee. Many objected to the ecological and environmental nature of his bulletins. His longest bulletin was never published. That bulletin was on the importance of songbirds for the farmer. While Henry Ford was funding support to save America's songbirds in Congress, Carver was part of the early environmentalist movement in the South. Carver argued the necessity of songbirds for insect control, which was not popular in the Southern states. This idea of the economic harmony of nature and the farm was a core one to Carver. He argued in over thirty pages the economic benefits of eighty birds. Associates at Tuskegee attacked some of his stated benefits of songs and cheerfulness. Henry Ford would have agreed strongly, but many at Tuskegee felt he had gone too far from the idea of useful hints for farmers. Carver also attacked the unnecessary leveling of forest and the need for conservation. Again, conversation at the time was not politically popular in the South. For these reasons, this bulletin was never published. Carver could even think outside the box, suggesting that herds of white-tailed deer be introduced in the area as a means of meat supply. Deer could feed off the nuts, fruit, and wild food that went to waste in the Cotton Belt. Carver, however, would strengthen his work to put nature study in schools. When he couldn't change the minds of the farmers, Carver turned to the education of their children to change the future. This concept of educating for future change was one of strong agreement between Carver and Henry Ford.

A later Carver bulletin showed farmers how to make white and color washes from the local clays. Carver argued that well-painted buildings and equipment improved farm efficiency and farmer pride. Carver went back to his own development of paints to show farmers how to make washes of blue and green by adding a little laundry bluing. He had a blue dye from rotten sweet potatoes for his washes. His washes went from pink and red to lavender and green. He added a permanency to these washes by adding boiled starch or rice. These washes were water based and treated the surface. Carver also developed a series of oil-based paints. He also supplied a recipe for one oil-based (linseed oil) paint that would last years.[9] Carver argued that farms should be neat, clean, well painted, and landscaped. One of his bulletins addressed landscaping. Carver suggested the use of local trees, shrubs, grasses, and flowers for landscaping (Bulletin 16: *Some Ornamental Plants of Macon County, Alabama*).

One of his most popular bulletins was the 1903 Bulletin 5, *Cow Peas,* which would be a blueprint for his future work. Carver argued that cowpeas (black-eyed peas) could replace the use of alfalfa as nitrogen-producing rotation planting. Cowpeas also offered a nutritious food as well. Carver included recipes and directions in using cow peas in

Tuskegee's first agricultural building where Carver originally held class in 1899 (Bentley Historical Library, University of Michigan, File HS8579).

soups, salads, puddings, and coffee. In total, the bulletin offered twenty-five recipes. He argued that cowpeas could be the "Boston beans of the South." He not only offered a Boston baked cowpeas recipe, but a new recipe for "Alabama baked beans." Carver also added a cowpeas and rice recipe for those favoring New Orleans–style cooking. If Carver were alive today he could make a living writing for magazines such as *Popular Science, Organic Gardening, Woman's Day, Good Health,* and *Scientific American.* His skill as a true science writer explains the worldwide demand for his bulletins. Carver's bulletins had the perfect balance of science and writing that could make the farmer interested in topics such as soil chemistry. He would eventually have his own syndicated newspaper column; he wrote a small book, *Economic Botany,* and a booklet for the scouts—*Nature Study Chemistry for Boy Scouts.* He was a collaborator for the *Nature Study Review* of the Teachers College of Columbia University. He continued to publish as a mycologist, identifying a number of new fungi.

Carver published a bulletin in 1906 on sweet potatoes that followed his earlier 1898 bulletin on sweet potatoes. The high yields and starch value of the plants made Carver a huge supporter of sweet potatoes years before peanuts. He noted that sweet potatoes would thrive in and improve the poor soil. Sweet potatoes had been the root of South America's oldest and greatest civilization, growing in the mountains where corn could not. Carver envisioned it as having the same potential for the New South. Furthermore, he showed that a farmer could get two crops in a single year of sweet potatoes. In 1910, Carver would issue a similar bulletin on sweet potatoes called *Possibilities of the Sweet Potato in Macon County*. Carver's experimentation over the years had increased the average yield from 37 bushels per acre to over 260 bushels per acre. Carver hoped to make the sweet potato a cash crop for farmers, but his first step was to show them its immediate uses as a food. Furthermore, the sweet potato offered an array of other uses.

Carver demonstrated that quality starch could be produced from sweet potatoes. Like Henry Ford, whose bakeries were making soybean flour bread for his employees, Carver was making sweet potato flour bread for the Tuskegee students. Sweet potato flour was one of a few that could be successfully mixed with wheat flour. Carver invented both a commercial method and a simple method for the farmer's wife. Carver's sweet potato flour and simple recipes made it into national magazines such as *Ladies' Home Journal*. Carver's success with the manufacture of sweet potato flour would bring him to Washington when the United States faced a critical shortage of flour in World War I. Because sweet potatoes stored poorly, he developed methods to dehydrate them. He noted that dehydrating could supply food for the hard winter months just as the farmers' ancestors had used them in the southern starvation days at the end of the Civil War. It is not surprising that in recent years NASA has used dehydrated sweet potatoes as a food source for our astronauts. Carver also developed a sweet syrup, vinegar, and alcohol from the sweet potato.

Besides its food value, Carver explored further uses. These would include glue, library paste, shoe polish, and rubber. He pioneered the use of sweet potato dyes, which also had a long history. Peruvians had long been using sweet potatoes to make an array of clothing color dyes. Carver discovered that sweet potatoes had a mixture of colors in their molecules, ranging from the obvious orange and yellow to an unexpected blue/purple. Carver was able to use acidity and citric acid to produce the full array from yellow to black. From rotten sweet potatoes, Carver extracted a deep blue dye. Carver dyes attracted a lot of attention in a world controlled by the German chemical dye monopoly.

Finally, in 1916, Carver published the cornerstone of his research bulletins with *How to Grow the Peanut and 105 Ways of Preparing It for Human Consumption*. Carver was constantly in search of nitrogen-producing crops to rotate in the poor southern soil, and crops that could supply protein to families having little meat in their diets. Carver even developed a type of food bar from sweet potato syrup and peanuts. While peanuts had been grown in Virginia since colonial times, they were a bit too novel yet to be adopted fully as a cash crop. Kellogg Company and H. J. Heinz were successfully marketing peanut butter in 1916, and peanuts were a popular snack at baseball games. Still, Carver realized he needed to develop more commercial uses, which was part of the focus of his bulletin.

In his 1903 bulletin, *Cowpeas*, Carver added a scientific discussion on the use of fertilizer by the famous German chemist Justus Von Liebig (1803–1873). Liebig was known in Germany as the "father of the fertilizer industry." Liebig was an expert in the new field of organic chemistry and was one of Carver's personal icons. Interestingly, Von Liebig was an idol of Henry Ford's as well. Liebig was known for his "law of the minimum," which described the effect of individual nutrients on crops. Carver had studied Liebig's research and summarized his work in fertilizers into the "four laws of Liebig" in his 1903 bulletin:

1. A soil can be termed fertilizer only if all the needed nutrients are present in the soil, and according to Liebig, the soil was as good as its least abundant nutrient (law of minimum).
2. A crop takes nutrients from the soil and only some are returned by the atmosphere (Carver was talking mainly nitrogen here) while others are lost. Liebig had been the first to discover the full role of nitrogen in the soil.
3. Fertilizer is required for full restoration of the soil nutrients.
4. Fertilizer from animal waste is not sufficient to fully replace the nutrients.

Carver proved to be a real student of Liebig in applying his laws to practical use. Liebig was a big supporter of crop rotation. Carver also improved on some of Liebig's earlier methods in chemical analysis. It had been the early work of Liebig that had interested Carver in the study of sweet potatoes and peanuts from his student days at Iowa State. Liebig would also be the inspiration for plant composting in the garden. Carver's studies and improvements on the work of Liebig should silence any criticism of Carver's lack of scientific methodology. For this early period of organic chemistry, Carver demonstrated an in-depth knowledge and the potential for applications of scientific theory.

By 1905, Carver added a weekly publication known as the *Messenger*. The bulletins, leaflets, and papers quickly won national praise. He struggled to get money to print them as the demand by the Department of Agriculture increased for reprints. Carver offered monthly meetings for farmers to discuss ways to improve and reduce costs. Carver constantly promoted the rotation of crops with soybeans and peanuts to improve the state's extremely poor soil. He trained farmers in making compost to replace expensive fertilizer. His popularity grew with farmers who were struggling against a boll weevil infestation of cotton crops. Carver got them interested in peanuts, sweet potatoes, and other legumes as an alternative crop as early as 1903. He even tested the use of sugar beets as a cash crop replacement for cotton.

Legumes and beans offered the greatest potential for the poor black farmer. It offered a replacement for cotton and a means to feed the family. They improved the soil through nitrogen fixation and offered protein on the level of steak. Through his manufacture of canned beans in the late 1800s, Heinz had brought high protein to the diet of millions of immigrant factory workers who couldn't afford to buy beef. Carver believed he could do the same thing for the families of poor southern farmers with cowpeas, peanuts, sweet potatoes, and soybeans. He wanted to free the South from cotton, which had ever-lowering economic value and depleted the soil of nutrients. The problem was the hold that cotton had on these southern farmers and the culture of the area.

5. Tuskegee and Detroit

Carver even experimented with silk as a replacement for the dependence on the cotton crop. James Wilson in the McKinley administration secured money for Carver to look into silk production for the South. Wilson also got 300 mulberry tree cuttings and silkworms sent to Carver. Mulberry trees were the food required for silkworms, and Carver had had some experience with mulberry trees at Iowa State. This effort eventually failed, but it was part of Carver's vision to replace King Cotton in the South. Carver believed that better farming would be the economic path to success for southern blacks. This was the type of work and research that Carver had been raised to do, going back to his thrifty foster father, Moses Carver. Carver was also able to get funds to establish a weather station for the Department of Agriculture, which opened the door to more research funding. With an experimental station, Carver could research better strains of weevil-resistant cotton, experiment with alternative crops, and, most importantly, educate the farmers.

As early as 1901, Carver experimented with soybeans from China as a nitrogen-bearing crop to rotate. Soybeans had little interest in America at the time. The problem was to find new uses, develop food recipes, and convince the farmer. In 1902, Carver discussed their possible uses far ahead of most agricultural experts, and he looked at improving their production. Carver even addressed potential problems of soybeans. On his first sample of soybeans in 1901, Carver identified the "frog leaf fungus," which years later would plague American soybean production. Over the next ten years, Carver would discover an array of soybean uses from oil to flour. In 1924, Carver had invented a dye from soybeans as well as one from his favorite dye-making plant—the sweet potato.[10] Decades later the soybean would be the source of conversation with his new friend, Henry Ford, who was also a devotee of the soybean. Both men would ultimately spend the last months of their lives working on the potential uses of soybeans in plastics, food, dyes, oils, and paints. Carver's research continued to amaze his colleagues at Tuskegee, local farmers, and a growing national following. Even Henry Ford, who subscribed to all the Department of Agriculture's bulletins, heard of Carver's initial work on soybeans and their potential uses in 1903.

Carver started his extensive research into soybean use in 1904 long before Henry Ford's interest. He had been growing them experimentally since 1896. In fact in 1901, he became the first to identify the fungus *Cercospora canescens* (frog leaf spot) growing on soybeans, which was a major discovery in mycology.[11] The discovery and ultimate control of this fungus would be critical in the development of the soybean as a cash crop. Soybeans offered the nitrogen properties that Carver, who had studied the nitrogen-fixing properties back in the 1890s, wanted to see in crop rotation. Benjamin Franklin had even experimented with soybeans, but real interest did not start until Commodore Perry brought back soybeans from Japan in 1854. An experimental station had tested them in New Jersey in the late 1870s. Carver had suggested them in 1897 in his Farmer's Institute, but farmers were hesitant to use them because they saw few uses. Carver's research then focused on developing potential uses. Carver was one of the earliest to separate the soy protein for use as a type of milk; this would later lead to his work with peanuts. In fact, farmers' unfamiliarity with soybeans moved him to shelve the work in favor of cowpeas and peanuts. Carver also realized that southern farmers needed a commercially viable crop in their rotation. These farmers could not afford any lost time in crop production. Carver's work did lead to a research program initiated by the Department of Agriculture in 1907.

Still, not everyone at Tuskegee saw Carver's work at the experimental station as having value. Unfortunately, Carver could not stop the badgering and undercutting by his faculty rival, George Bridgeforth. Bridgeforth had won over Washington who was tired of Carver's whining and complaining about funding and administrative reports. Carver lacked administrative skills, yet Bridgeforth excelled in them. Bridgeforth proved adept at the report writing that Washington required. Washington implemented an awkward structural solution between Carver and Bridgeforth. Carver was to become director of experimental research and the agricultural station, while Bridgeforth would be in charge of the teaching side of the department. But even this solution failed to solve all the fighting.

The battles continued, with Carver threatening to resign a number of times. Only Providence seems to have kept Carver at Tuskegee. For his part, Carver believed in mysticism and divine destiny, which probably explains his perseverance in the administration opposition he faced. In this respect he was convinced that the educating of poor blacks was a God-given mission of his.[12] Like so many achievers, Carver believed he had an internal calling that prevented him from despairing, made him believe the criticisms were wrong, and ignited a desire to overcome. Carver was determined to show the importance of his work, so he looked at King Cotton, which ruled the area. He turned to research as his relief, increasing the breadth and intensity of his experiments.

Some of the earliest experiments at Carver's station were to improve the yield of cotton crops, which remained the major crop of the Black Belt. Carver had initially been interested in alternatives to cotton but felt the resistance of farmers to change from cotton. However, Booker T. Washington pushed to do more with cotton experimentation; so in 1903, Carver added sixteen acres to the experimental station for cotton experimentation. The area preferred the "bumblebee" cotton plant that yielded two bolls per stalk. Carver's experiments were highly successful. He mixed durable and high-yielding varieties to develop his own hybrids. Using hand cross-fertilization, he developed four very useful hybrids of cotton, yielding up to four bolls per stalk. Carver also experimented with the long-fiber-staple varieties needed to make strong rubber tires, which was a growing market that was being filled by imported Egyptian cotton. Carver also developed many products reinforced by high-strength long-staple cotton. These products included floor tiles and road-building blocks.

Carver used homemade fertilizer to further increase cotton yields which amazed local farmers who couldn't match his results. Carver suggested the use of sawdust, wood chips, acorns, beechnuts, and hay in animal pens to make fertilizer. He suggested leaf collection in the fall to build compost piles. Since cotton demanded huge quantities of compost, Carver promoted the use of swamp muck, which was available. There was very little that Carver did not see as a potential fertilizer. Carver even argued for the use of "night soil" (a euphemism for human waste) mixed and properly treated as a great fertilizer.

The Tuskegee Institute experimental station played a key role for the poor farmers, both black and white, on soil improvement. The Morrell Act and the Alabama Department of Agriculture funded the official experimental stations (of which Tuskegee was not one) through the sale of commercial fertilizer. Commercial fertilizer carried a quarantine tag from the state that accounted for one-third of the funding for experimental stations. Therefore, it was to the advantage of the state and its experimental stations to promote commercial fertilizer sales.[13] This was an example of how sometimes government accidentally worked against the people it was trying to help. The problem was that poor farmers could not afford the additional costs of two to three cents a pound on cotton that sold for four cents a pound. Carver's work with composting offered an alternative to the high cost of commercial fertilizer. He demonstrated that composting and field rotation was superior to the use of commercial fertilizer. Composting and manure offered a synergy to Carver's belief in the diversity of the family farm.

His visible results at the experimental station went a long way to win over the hearts and minds of local farmers. He wrote several bulletins to educate farmers on crossbreeding and fertilizer production. He showed that crop rotation, fertilizer, and crossbreeding were valuable tools for farmers. His work caught national and international attention through the distribution of his bulletins. Carver had done his master's thesis on Luther Burbank's hybridization work and applied it to cotton with great success. Cross breeding plants would be a lifelong hobby for Carver. Throughout his life, he crossbred amaryllis and lilies. His better-known work in this area was with cotton, however.

The German government worked with Carver to introduce these hybrids into Africa for cotton production. The "Carver hybrid" was also successfully planted in Australia. Carver's hybrids improved oil-bearing cottonseeds. Carver also worked on uses for cottonseed oil, which was considered a waste product. Cottonseed oil was excellent for cooking and paint making. Over the years, Carver would even work with a cotton hybrid known as "bald-headed" cotton that could be raised for seed versus fiber. Still, like so many of Carver's ideas, it had to wait for the development of more powerful seed presses. Carver would be pulled off his cotton research as the boll weevil infestation was slowing progressing northward and eastward at sixty miles a year from Mexico, devastating cotton fields throughout the South.

Carver broadened his personal interests into the area of education since future teachers were a major part of Tuskegee's mission. Early on in his career at Tuskegee, Carver became involved with the grade schools in Alabama. In 1902, Carver published a bulletin, *Nature Study*, which became popular with teachers. The next summer he traveled to Knoxville, Tennessee, to give a summer nature study course. The trip allowed Carver to once again find time for painting flowers, plants, and landscapes. He taught the children how to make their own paints from nature. On his return to Tuskegee, he set up a children's program for nature study and gardening. Some nearby grade schools that adopted Carver's methods became highlighted as model schools by the federal government and brought in more funding for Tuskegee. His belief that children should be taught basic gardening and nature study in grade school was shared with Henry Ford and the earlier teaching methodology of William McGuffey.

In these early years at Tuskegee, Carver continued his academic research during the summer when students were on vacation. Mycology was a field that Carver excelled in because it still required the Victorian methodology of collecting, identifying, and classifying. Carver's list of southern fungi he identified was published in the *Journal of Mycology* in 1902. Two of the new species listed would be named *Metasphaeria Carveri* and *Taphrina Carveri* in his honor. These two fungi attacked maple trees. Carver had identified many crop fungi earlier on in the developing science of mycology. He was the first to identify several fungi that attacked soybeans. Carver's leadership in the science of mycology is often overlooked, but he was truly a national expert in the field.

Carver cooperated with the Alabama Polytechnic Institute in a monograph of 1,120 species of fungi in the State of Alabama in the 1920s. His fellow author, Franklin S. Earle, later became the mycologist at New York Botanical Gardens. Carver would often exchange samples with the Botanical Gardens over the years. Carver's work in mycology from 1899 to 1910 showed how technical his early career was, and how many have overlooked it. He corresponded with the best mycologists in the nation on technical topics and the scientific classification of rare fungi. His work in mycology continued throughout his career, and in 1935, he was named collaborator for the Department of Agriculture's Mycology Disease Survey. He worked with the Department of Agriculture on the identification of grasses in the United States. He also worked with the Smithsonian Institution on cataloging medicinal flora.

Probably the brightest part of Carver's early years at Tuskegee was the potential national and commercial recognition of his research. Carver was continuing to invent and promote commercial products such as a metal polish, insect poisons, and laundry bluing agents. While he failed to fully commercialize them, he used them to save thousands of dollars at Tuskegee in laundry, cleaning, painting, and food bills. In his bulletin, *Pickling and Curing of Meat in Hot Weather*, Carver worked with the Armour meat packaging plant in Montgomery, Alabama. Carver was focused particularly on the storing of pork, which was difficult in the hot weather of the South. Pork was easy meat for poor farmers to raise, so Carver tackled the problem of preserving and storing. He failed to interest any commercial meatpackers, but he did convert Tuskegee and local farmers to using pickled pork versus buying canned meat. Pickled pork became very popular in the Black Belt as smoked ham was in Virginia. Hog raising increased, and farmers used Carver's suggestion of using acorns and rotten fruit as feed to greatly reduce cost.

Carver also gained national interest from the popular press during the pure food movement of the early 1900s. Corporate fraud and labeling was a big issue in the general movement. Carver had become known in Alabama for his work with analyzing samples from farmers for fraud. In 1905, in the battle for the Pure Food and Drug Act of 1906, Carver related one problem in *Collier's Weekly* of commercial fraud. Carver said, "You wonder why your cattle are not satisfied with their food, why they fail to yield the milk and butter which you have the right to expect?"[14] Carver noted from his work with samples of feed that some had been adulterated by nearly 50 percent with sawdust and corncobs. Carver became popular in this type of society magazine which focused on issues of corporate or government problems. Booker T. Washington promoted a lot of this national fame for Carver with his strong ties in the eastern press and society. Every mention of Carver was a mention of the Tuskegee Institute. Booker T. Washington was the master at creating a myth. These early articles were a gateway to promoting commercial ideas as well. Still, commercial support never developed for most of Carver's ideas.

Carver found some interest by paint manufacturers. His clay-based pigments of carmine and Prussian blue had attracted some commercial developers in 1902. The story of Carver paints was featured in local and Atlanta newspapers, and a group of investors formed a paint company to use the clay deposits. Carver appears to have found that by using various levels of oxidation of the local clays, he could produce an array of colors. Experimenting with oxidation levels was a strong piece of chemical research, which attracted other scientists in the field. He had also discovered a rare pocket of blue clay around Montgomery that could be oxidized to a deep blue purple. The paint company failed for reasons that were not clear, but Carver would be inspired to look for more commercial products.

Years later he would use the clay to produce a red paint for cars that did find commercial use at General Motors, although Carver never got any financial rewards from it. For Carver, it was a matter of not only capitalistic gain but professional recognition. Professional recognition would prove elusive even as he gained the support of white newspapers. There was far more than racism at the heart of the issue. There was a real bias from university professors to Carver as a scientist. Carver, like Edison and Ford, was the last of a Victorian breed of scientist held in low esteem by the twentieth-century university scientists. Carver with botany and Edison with chemistry had often questioned the academic classifications of plants and chemicals for more pragmatic ones. Their experimental methodology versus a theory-based one was considered less sophisticated. Carver's strong religious views were also problematic. Carver drew similar criticisms of his Bible studies from university theologians, but again the results of packed noncompulsory classes had to silence many.

Carver's Bible studies showed the same creativity as his research. He often dressed up as the characters to be studied, taking questions from the class. His creation studies were particularly popular, with Carver tying science to the Bible. He used specimens to argue for the creation story of the Bible. When studying Moses and the exodus, he brought in samples of potential manna from the woods. His classes always inspired lively discussion. His science and Bible studies quickly won over the students at Tuskegee. Carver was an active supporter of Tuskegee's largest student organization, the Young Men's Christian Association (YMCA). Carver's Saturdays and Sundays were filled with Bible studies, church, religious meetings, and YMCA meetings.

In 1906, Tuskegee had clearly made the national headlines. Andrew Carnegie visited to see the library he had donated years before. John D. Rockefeller had also visited and donated buildings. The institute made a pair of golf shoes for Carnegie as a gift. William Taft, the secretary of war, visited and spent time with Carver. Many educators such as the president of Harvard University visited to look at Tuskegee's blend of manual and academic training. While most of this interest can be attributed to Booker T. Washington's promotion, Carver had strong ties in the McKinley and Roosevelt administrations' Department of Agriculture. Still, Carver was having problems finding his place at Tuskegee.

Carver would also have a strong opinion of the education of the youth, which he

Tuskegee's second agricultural building in 1918 (Bentley Historical Library, University of Michigan, File HS8580).

shared with Henry Ford. Tuskegee was, after all, a school to train teachers. He not only promoted nature study and gardening but suggested the linkage to crafts study. Carver, like Ford, shared a love of the McGuffey approach to elementary school. The *McGuffey Readers* had always devoted many pages to nature study as a means of inspiring future scientists and engineers. Nature study could serve as a gateway to science study as it had with Henry Ford. Carver published a number of nature study and science leaflets for schoolchildren, which also proved popular. Carver added his own twist to the elementary curriculum with his addition of a school garden. His 1910 leaflet, *Nature Study and Gardening for Rural Schools*, is really a classic for educators in many ways. It blended an array of disciplines such as botany, chemistry, farming, and geology. It perfectly augmented the earlier work of William McGuffey in elementary education. It also trained elementary students in the fundamentals of science and its use in everyday life. Carver became very popular with rural school principals, who often called on him for help.

Still, Carver struggled to meet the expectations and demands of Booker T. Washington in teaching and the administration of teaching. And Washington never fully fulfilled his promise to supply Carver a first-class laboratory. Carver had to make his equipment out of trash and broken tableware. For chemicals such as zinc sulfate, he scraped zinc off lids. The poor conditions of the laboratory often hurt Carver in his pursuit of commercial applications. He continued to contribute to major savings at Tuskegee with his manufacture of paints, stains, and fertilizers. Tuskegee Institute with all its publicity had an ambitious program that taxed its funding, often leaving Carver's

5. Tuskegee and Detroit

lab needs unfulfilled. Both Carver and Tuskegee would adapt and grow on the national scene.

While Carver was getting off to a rocky start at Tuskegee Institute, Ford was hoping for a rocky start in the automobile business. Ford's 1896 Quadricycle demonstration had hardly started a career. He did start to gain a reputation, though, as his strange vehicle became common on the streets of Detroit. But it was not up to the commercial competition of the time. Ford went to work on the production of a second car. It would take a year to build, and it would take three years to line up investors to launch a manufacturing company. Finally, investors such as prominent Detroit businessman William Murphy; the mayor of Detroit, William Maybury; George Peck, president of Edison Illuminating; U.S. senator Thomas Palmer; and others invested in Ford. On August 5, 1899, the Detroit Automobile Company was incorporated at a capitalization of $150,000. This hopeful start ended in January 1901 as Ford failed to move from prototype to production and the company's losses mounted to $86,000. The same year rival Oldsmobile had profitably produced 435 cars at a selling price of around $2,000. In the luxury-class auto, Alexander Winton produced over 300 cars that year.

However, it was far from an end for Ford. His most important investors stayed with him in the development of a car. The operation continued in the Case Avenue closed factory. Ford and his investors realized he was losing ground as automobile companies were starting by the hundreds. Ford needed more national attention and a bigger name, and he had been working on a new project for months. Henry Ford was working on a race car, with his eyes to Michigan's first auto race to be held October 10, 1901, at the Grosse Point racetrack. It would be a daylong event with the world's greatest auto racer and manufacturer, Alexander Winton, coming to Detroit with his racer.

This race would be reflective of the state of the automobile industry in 1901. The series of races at Grosse Pointe would follow the breakdown of the industry. There would be preliminary races for electric cars, steam cars, and gasoline cars, to be concluded with a ten-mile main event. Electric battery power vehicles and steam cars were more popular at the time. Electrics were slow and limited in range. Steam took time to start up. Electrics appeared to be the popular car of America's great inventors such as George Westinghouse, Thomas Edison, and Michael Owens. Gasoline offered the best power system for speed, and the largest gasoline car producer was Alexander Winton's of Cleveland. Ford believed gasoline to have the quick start and speed to replace the horse. Winton's success was based on his winning races across the nation. It was a business model Ford wanted to emulate. Winton was producing 300 high-end cars a year with famous owners such as Andrew Carnegie. Winton cars were heavy (3,000-plus pounds) with powerful engines and Goodyear tires. Henry Ford's vision was unique in that he wanted to bring the automobile to the masses. Most cars of the time could only be afforded by the rich.

Ford was of little interest to the race's draw that day; it was the great Alexander Winton. But Winton and his racer were not the only draw. The French-built "Red Devil" car, the world record holder with Michelin tires, was to be there also. This French-built car had its celebrity following in the families of Vanderbilt of New York and H. J. Heinz of Pittsburgh. The Red Devil held the world mile record at one minute and fourteen seconds. The Red Devil came with an American price of over $22,000, an amazing sum for the time. The Detroit Racing Club was offering a $1,000 prize and a hand-cut Libbey Glass punch bowl. Ford could use the $1,000, having already poured over $5,000 into the development of his racer. The crowd would reach over 8,000 on a beautiful autumn day.

After the preliminary races of the day, the final race boiled down to Ford and Alexander Winton. Most of the other cars suffered mechanical problems prior to the race. Earlier in the day, Winton had made headlines by breaking the world speed record for the mile. Ford had an underpowered but much lighter car. In fact, Ford had designed a car far lighter than most built in America. He hoped his lean design would help bring down the overall cost. His mechanic, Spider Huff, would be riding with him to balance the car on the turns. Winton ran ahead throughout the early going of the race but in the final lap developed engine problems, allowing Ford to overtake Winton in the stretch. It was a great victory and one that fortified old investors and brought in new investors. It allowed Ford to incorporate once again.

A new company, Henry Ford Company, was formed on November 30, 1901, but Ford's share was a mere one-sixth. Ford was never comfortable, as William Murphy was clearly running the show. Murphy was disappointed with Ford's slow process towards a commercial car. By 1901, there were hundreds of carmakers in the United States. Murphy brought in his own mechanical expert, Henry Leland, who was producing motors for the successful automobile maker Ransom Olds. Interestingly, Henry Leland would found Lincoln Motor Company, which would later be purchased by Ford in 1918. The vision of the new company was clearly that of Ford with a goal of a $1,000 car. The best-selling cars of the time were the Oldsmobile and the Winton, priced around $2,000. Still, Ford's hands were tied in making the changes he felt were needed. He believed the car's weight had to be reduced, which was contrary to the design of the period.

Unfortunately, Ford found little agreement with the large investors and would leave the company in March of 1902. Ford had continued to improve on his race cars. Ford's "999" would enter and win a number of races in 1902. The 999 would capture several speed records, putting the Ford name in the forefront. Ford, however, had to sell the car to pay suppliers. The notoriety brought Ford a new investor in Detroit's coal baron, Alexander Malcolmson. Malcolmson backed the formation of Ford's third company, the Ford Motor Company, on June 16, 1903. Malcolmson brought in James Couzens to manage the business and financial side, while Ford was general manager (of operations) with a salary of $3,000 a year. Arrangements were made with the Dodge Brothers Machine Shop to supply engines, transmissions, and axles. Ford set up an assembly plant on Detroit's Mack Avenue. He hired a dozen workers at $1.50 a day to assemble cars known as the Model A.

Ford produced a "standardized" car versus the customized building of manufacturers such as Alexander Winton. His standardized part sizes were possible. The Mack Avenue factory was far from an assembly line, but simple standardization kept costs down. Ford's approach would prove revolutionary and helped him achieve his goal of an affordable car for the masses. Cars were built in stalls by a group of workers capable of making ten cars in a day. Ford's costs to manufacture the Model A were $554 with a selling price of $750. Ford also offered some options in upholstering that added to his profit. The Model A offered a low cost with great quality, and sales took off. In his first eight months, Ford produced 658 automobiles with a profit of $150 per car. By the end of 1903, Ford employed 300 workers making 25 cars a day.[15] Success brought the internal struggles of Ford and the money people.

The pre–Ford Model T gasoline automobile industry consisted of two segments — the expensive luxury cars and the "big" carriage type cars. The luxury models ran over $5,000 and were names like Winton (of Cleveland), Pierce-Arrow, Packard, Pope (Toledo), White, and Duryea. The standard cars were the Buick, Studebaker, Overland, Oldsmobile, and Cadillac, which sold for $750 to $3,000. Electric still made up 50 percent of the market, were slower, and had a limited range. Henry Ford even considered building an electric car. Because he hated gasoline fumes, he toyed with the idea of an electric car for years.

Malcolmson and his backers were happy to continue making the Model A, but Ford still dreamed of a lower-priced car for the masses. The Model A would compete in price with the standard competition. Ford wanted to push further weight reductions and material costs. He believed he had to get the cost under $500 to open up the market. The average yearly wage at the time was $600, and Ford believed it was necessary to get the price well under that. Finally in July of 1906, Ford took control of his destiny by buying out Malcolmson. That would be something that George Washington Carver would never fully achieve; but in the end, both men would live their dreams.

6

The Model T and the Jesup Wagon

"Since new developments are the products of a creative mind, we must therefore stimulate and encourage that type of mind in every way possible."
— George Washington Carver

In 1906, both Carver and Ford were pursuing their visions with notable success and on the verge of major breakthroughs. For Carver, the breakthrough would be his famous Jesup Wagon that brought lessons to farmers throughout the South and would be the model for federal farm extension programs. The bulletins Carver distributed would be asked for around the world. The Tuskegee Institute had grown to 1,500 students and 156 buildings by 1906. For Ford, it would be his Model T that would revolutionize transportation and the American industry. However, the Jesup Wagon would be the equivalent of the Model T in agriculture. Both would change how people lived and the society in which they lived. Both would represent the mission of these two men.

While the internal struggles between Booker T. Washington and George Washington Carver seemed endless, they did team up on a breakthrough concept of a traveling wagon to educate the poor farmers of the area. There would be a struggle later on about whose idea it was, but history supports a confluence of the two. Washington had always been interested in offering help for Alabama farmers in some form. In the 1880s, Booker T. Washington had traveled on horseback to farmers to help them develop tenant farms. Washington had started the annual Tuskegee Negro Conference for farmer education in 1892. The first had 400 farmers; with Carver's fame growing, the 1898 Conference had over 2,000 attendees. Carver also launched a "Monthly Farmers Institute," but these poor farmers lacked transportation and travel time to get to the school, which meant loss of production time. The idea of farmer education had been a key part of the U.S. Department of Agriculture's support for Tuskegee's funding of an experimental station. With Carver's old professor and friend as secretary of agriculture, educational funding was increased.

Carver brought similar ideas of farm extension programs from Iowa State College, which had also held conferences and institutes. Carver had not only helped with the conference but also developed a four-week "Short Course in Agriculture" for training farmers. The short course was designed for winter when the farmers had time for education. Carver added courses for the wives such as cooking, sewing, and food preservation. Because of the duration of the conference, Carver arranged for board and lodging nearby at $2.50 a week. These short courses had a slow start of only 20 farmers, but by 1912, using the

6. The Model T and the Jesup Wagon

Jesup Wagon to publicize, they had over 1,500 enrolled. Carver also got the Department of Agriculture to supply free seed packets to help these farmers build productive gardens. It was clear that Tuskegee had now become Carver's long-term home. His passion for educating black farmers stemmed from his belief that this mission was given to him from God. This belief carried him past school politics, lack of funding, racism, and government resistance.

Carver had been amazed from his first days in Alabama at the lack of gardens. The climate of Alabama allowed for almost year-round food production in a simple home garden. Farmers were actually buying food and vegetables from the market. In many cases, these tenant farmers or sharecroppers were forced not to waste land on vegetables instead of cotton. Carver soon realized that it was an educational problem. Owners' demands required the whole family to work long days in the field, reducing real productivity. The sharecropper and the owner's farm would be more productive if a little agrarian capitalism was injected. Even in the days of slavery, more enlightened owners went to the task system, allowing slaves to own their house, own gardens, hunt, and have family time. Carver argued from the point that the owners could understand — more productivity.

Carver not only had to win over the owners but the farmers themselves. He used his bulletins and courses not only to promote gardening but also to teach their use in the preparation of meals. Carver promoted the family as an economic unit. He addressed the housewife in his bulletin *Three Delicious Meals Every Day for the Farmer*. Carver even addressed wild foods in his *Some Choice Wild Vegetables That Make Fine Foods*. He had to educate the whole family if he was going to change things. Carver was looking to the next level of freedom from slavery.

Carver first looked at changing the basic approach to child education. He had to change minds, and this required a change in the schools. Men like Carver, Washington, and Ford realized that real change in society comes only through the education of the youth. Black farmers were not only resistant to change, but the white landowners wanted to allow few extras that might take away from field time. The change in education would have to address both groups. Carver argued that nature study and gardening had to be added to school curriculum, and in this, he even faced resistance from schoolteachers. Carver developed an outreach-training program for common school teachers as well as the future teachers at Tuskegee. This program, established in 1901, was housed under the Bureau of Nature Study for Schools. He prepared a number of guides and outlines for incorporating nature study into rural schools. Carver worked with Booker T. Washington to gain donations for a traveling library. This approach was common in McGuffey-style schools of the Midwest. Carver believed that lack of understanding nature and ecological relationships was at the core of poor farm management. Carver was convinced that in the South it was possible for farmers to completely live off the land. Besides the commercial plants of the farm, wild vegetables, herbs, nuts, and fruits were readily available.

Carver prepared a textbook titled *Suggested Outline for the Study of Economic Plant Life for Use in Common Schools, High Schools and Academics*. In addition, he published a number of leaflets on nature for teachers on topics such as how to construct a hotbed winter garden. Carver's efforts became part of a larger national movement represented by the *National Nature-Study Review*, of which Carver served on the board. Carver was able to win over Booker T. Washington on nature study, and gardening for youth was a key to improving the lot of black farmers. Carver would recommend field trips for things like leaf identification, which William McGuffey had been a supporter of in the 1850s and 1860s. Like McGuffey, Carver promoted the planting of various species of trees, shrubs, and plants around schools to facilitate nature study.[1] Carver, in particular, promoted the planting of commercial nut and fruit trees. These commercial trees could have a long-lasting economic benefit to the black farmer.

Carver believed the problem was educational because of the long growing season available in the South. George Washington Carver took on the mission of helping the poor black farmers, who suffered from much inefficiency in their farming. He started by trying to wean them off cotton by converting to sweet potatoes, cowpeas, and peanuts. He hoped to expand their use of fruit through his earlier research and study of Luther Burbank's hybrid technology. Unfortunately, southern farmers used plums and peaches for hog feed because of the problems in preserving. "Canning" required jars and sugar, which made it too expensive as a preserving method for black farmers. Carver offered them techniques in dried fruit to expand their diet, publishing a bulletin, *How to Dry Fruits and Vegetables.* Drying fruit offered the perfect storage method for the South. Carver wrote of an old drying technique used in southern Europe, which he called "fruit leathers." These were really a sweet natural candy. Carver used overripe fruit to make a pulp and then rolled it out to cut and dry. These products are becoming popular today.

Still, the black farmer also had to struggle with poor soil from decades of cotton farming, and the cost of fertilizer was beyond the budget of most black farmers. He aimed at reducing costs with homemade fertilizer, and taught home economics to improve the family budget. In 1906, Carver published *Bulletin 6: How to Build Up Worn Out Soils.* His bulletins became extremely popular even with the small white farmers, which helped Carver overcome some suspicion of the white landlord farmers. Carver believed fully in homemade compost fertilizer because manufactured fertilizer was extremely expensive. Earlier on, Carver had followed the early nineteenth century German chemist, Justus von Liebig, whom he referred to often. His former teacher, Henry C. Wallace, had also been a major advocate of soil development. Carver added components of his composting program to his suggestion for school nature study. He became passionate about analyzing the soil chemistry to adjust the right fertilizer. Carver was constantly experimenting with different types of wastes to improve the soil. Another of Carver's beliefs, which Henry Ford would also agree with, was to plow very deep. Both men often preached that farmers didn't go deep enough in turning over soil.

Booker T. Washington had encouraged Carver in his writing of bulletins and educational outreach. Carver had a very strong background in farmer bulletins and pamphlets, and it was a real strength. Professor Wallace's father had published the magazine *Wallace's Farmer,* which Carver contributed to and read. The magazine was extremely popular in Iowa. The publication remains to this day. *Wallace's Farmer* was a mix of an almanac and folksy technology. Even religious and child-oriented content were added. It addressed farming themes such as:

6. The Model T and the Jesup Wagon

- How to plow soil
- How to get nitrogen into the soil
- Composting
- Fattening geese and chickens
- Using clover to improve the soil and feed cattle

These themes are very similar to those in Carver's bulletins and newsletters at Tuskegee. *Wallace's Farmer* would, over the years, be edited by future secretaries of agriculture, William C. Wallace and William A. Wallace.

One problem was the high number of illiterate farmers in the Alabama "Black Belt."[2] For his part, Carver had been making weekend trips to churches and farming groups to talk about ways to improve. Carver also expanded on Washington's Monthly Farm Institute by taking in the communities and the Macon County Fair. He had also been asked to set up exhibits at county fairs throughout the state. Carver also initiated a Farmers' Institute Fair at Tuskegee every year. These Farmers' Fairs became popular with black farmers who felt more comfortable competing for prizes among other blacks. The idea of black farm fairs at the county level spread rapidly throughout the South. Carver's advice was so popular that it offered a type of integration of white and black farmers, who both suffered from the poor soil of Alabama. Carver got his students involved in this outreach as well. Eventually, he would head up a committee to develop a traveling wagon and school.

The idea of a movable school to help reach black farmers fit the vision of both Washington and Carver. It was the perfect meeting of Washington's ambition with Carver's skills. Washington and Carver also realized that success would mean more government funding and private philanthropy. The South, and particularly the Black Belt, was suffering in a true economic depression that had roots going back to the Civil War. The Civil War had freed the slaves, but for the most part these former slaves were struggling tenant farmers. The white economic infrastructure was no better off with the decline of King Cotton.

It was estimated that there were 2 million black farmers in the South, most of which were illiterate and had no means of transportation. In addition, the success of these tenant farmers was important to the white landholders who needed the rent from their crops. Farmers were struggling, paying as much as $1.50 an acre in fertilizer to make the soil usable, all while paying $2.50 an acre rent. This was not sustainable. Many of these tenant farmers had debts over $1,000 from paying rent and fertilizing their land. Their debt was also becoming the debt of the white landholders. Booker T. Washington had reported in the 1890s that 80 percent of the black farmers lived in one-room cabins and mortgaged their crops for food to live off of. The whole tenant farming system was near collapse, and the black farmers had exchanged iron chains for economic chains.

Carver had developed the techniques for them to make their own fertilizer, but he needed a means of taking the information to these tenant farmers who often could not read. Carver was anxious to get this information to the black farmers, but their lack of reading skills limited the use of bulletins. A movable wagon or school made more sense, and Carver was aware of the "Seed Corn Gospel Trains" of Iowa State, which took lectures and demonstrations to farm-area train stations. He was also aware of the use of "movable agricultural schools" in Europe. It was an idea he believed he could sell to the Department of Agriculture. Booker T. Washington could also sell the idea to his northern backers.

The idea of a movable school fit perfectly Booker T. Washington's vision of black farmer education. Carver came up with a design for a horse-drawn wagon with large charts, samples of soil and products, and a milk separator for cheese, different types of plows, milk-testing equipment, soil-testing equipment, and a cotton chopper. The estimated cost of $567 would require some financial aid. Getting money for such projects was where Washington excelled. He found support from New York banker and philanthropist Morris K. Jesup. In 1906, the wagon known as the "Farmers' College on Wheels" and the "Jesup Agricultural Wagon" started going to farmers. The wagon coordinated meeting places to include both white plantation farmers and poor black tenant farmers. A meeting of white and black farmers for education was a major breakthrough in itself. Success came quickly, attracting over 2,000 a month in 1906 to these meetings. Part of the success was Carver's program of offering an integrated approach to black farmer education. The training offered cooking lessons, recipes, the manufacture of cleaning products and house paint, and general tips on saving money.

The Jesup Wagon caught the attention of the Teddy Roosevelt administration and Carver's friend, James Wilson, who was still secretary of agriculture. Roosevelt had visited Tuskegee as vice president with President McKinley. James Wilson had written a policy report that southern farmers faced disaster from the boll weevil but had found that education of the farmers was needed. Wilson supported funding the Jesup Wagon as a means to educate the farmers. One problem in obtaining federal money was that black experimental stations could not directly be funded.

In May of 1906, the Jesup Wagon got its start with George Bridgeforth as the operator. The slow travel of a mule-drawn wagon required a full-time operator. After a few months, the idea caught the attention of the Department of Agriculture. Booker T. Washington and Carver quickly exploited the interest. Washington was a master of publicity and had strong support in the North. The publicity pushed the government to take action. The result in November of 1906 was the naming of a "special agent in charge of the Farmer's Cooperative Demonstration Work." The financial support of the federal government was, however, marginal. The federal government supplied one dollar for the salary of the agent, ten dollars in travel expenses, and free mailing. The General Education Board of New York and Tuskegee supplied $1,090 for the salary and expenses. The Department of Agriculture had argued for a white agent, but Carver was successful in getting his black graduate student, Thomas Campbell. Booker T. Washington and Carver achieved federal recognition of a black agent.

The Jesup Wagon would be hard but rewarding work for Thomas Campbell. It was a life of dusty, hot roads with little traveling comforts. In 1907, Campbell had traveled 800 miles within Macon County, visiting 100 farms and giving demonstrations.[3] Within a few months, the Department of Agriculture had hired another black graduate of Tuskegee, John Pierce, to be an agent in Virginia. Thomas Campbell would prove as successful as his patrons, working for the Department of Agriculture and ultimately supervising two agents in seven southern states. The advance of black agents and the southern extension program came through the hard work of Washington and Carver in the face of the white bias of the Agricultural Department. Most of the Department of Agriculture funding was sent to the state-controlled universities, leaving Tuskegee to compete for private funds and small government grants. But even with less funding, Carver's expertise was more highly prized in the South than any white agent.

Besides the Jesup Wagon, Carver continued and expanded his farm extension program via county and state fairs. He moved into commercial possibilities for both poor whites and blacks. Carver had become very artistic at building exhibits, and he was popular with fair planners throughout the South. At the Alabama State Exposition in 1911, he had the most popular exhibit. It featured an array of colored paints and washes made from Alabama clays. Alabama's clay deposits offered rich red and yellow colors that had been used by the Indians for painting. He highlighted the use of other state minerals such as azurite (copper carbonate) and iron oxide as a coloring agent and used the mineral betonies to de-ink newspaper and make it useful. He showed that Alabama white clays of fine kaolin could be used to produce fine ceramic "china." He displayed an array of porcelain and stone pottery. Part of the exhibit showed how fine paper could be made from vegetable fiber. He showed a wood stain that could be produced for seventy-five cents a gallon from peanuts versus four dollars a gallon for commercially available stains. This would allow cheaper southern pine and spruce to be used in furniture. Carver produced a yellow oak and dark walnut stain for wood from peanuts. He demonstrated the high yield and endless uses of sweet potatoes. These exhibits were starting to be in demand around the country, and state governments wanted the federal government to help.

The idea of an extension program would become an important part of the Department of Agriculture, and the initiative expanded into conservation and land management. The success of the program demanded national approaches, which led to the passage of the Smith-Lever Cooperative Extension Act of 1914. Congressional bias, however, would not reward Tuskegee's success in pioneering the program. Ironically, the act actually took money away from Tuskegee because it was to be administrated by state-run institutions of the 1862 Land Grant Act. It did allow for a separate Negro extension under the supervision of a land-grant institution. The work of the Jesup Wagon of Tuskegee came under the direction of Auburn University. In 1912, Auburn received $70,000 in federal funds for agriculture extension while Tuskegee received only $1,500. Still, the concept of agricultural extension, which Carver had pioneered, became the cornerstone of the Department of Agriculture. Carver's Jesup Wagon and the extension program was the first major step in Carver's life mission.

Ford had one mission in his early effort — that mission was a lean-engineered car for the masses. Mass production would be the result of his quest to achieve this mission. On New Year's Day 1906, Ford announced he would offer a car for under $500 to fulfill his goal. Ford's car for the masses, the Model N, was launched in the summer of 1906. The lesser-known Model N was to be the prototype for the famous Model T and the Ford assembly-line process. This model N was being built at the new Piquette Avenue plant. The Model N would replace the Model B at $2,000 and the Model C at $950.[4] The

Model N would capture a larger market with its lower selling price. Ford's sales success was not just in his selling price, but in driving down costs and improving quality. His automobiles were lighter and more rugged than the competition. His cars got more than twenty miles per gallon and with a ten-gallon tank had a cruising range of 200 miles in an era of limited fueling availability. Ford was in a position to make a profit of over $55 per car with the Model N even before he fully automated his assembly. With Ford's car sales came new personal wealth.

Wealth would have a major impact on Ford, but more important was that his organization was now strong enough to free him of the day-to-day management. While Ford went to the office early each morning, his team allowed him some time for other interests in the afternoon. Ford liked this freedom, which he had first found in his chief engineer position at Edison Electric. Ford built an organization for the future. James Couzens was the company accountant and purchasing agent, Harold Wills was chief engineer, Clarence Avery was a cost-cutting production manager, and he had a young operations man in Charles Sorensen. The quartet of Couzens, Wills, Sorenson, and Avery could take Ford's ideas and turn them into reality. Couzens would become a multimillionaire from Model T profits and retire into politics. Likewise, Wills became a multimillionaire from his metallurgical patents for Ford Motor Company. Clarence Avery would enter the Automotive Hall of Fame as the designer of the first assembly line. Sorenson would win much fame as executive vice president of Ford Motor, and later in life he took over Willys-Overland Motors.

Ford had also developed a strong supply chain including the engine sand drivetrains from the Dodge Brothers and the bodies from the company of Fred Fisher (later Fisher Body). Ford forged a lasting relationship with the salesman that supplied his first racing tires — Harvey Firestone. Firestone's invention in 1907 of an easy rim system for replacing tires helped to popularize the Model T with the general public. Ford used blueprints to standardize his parts from suppliers. Ford's alloy steel development would give rise to Republic Steel. Ford developed his suppliers as his company grew. A strong supply chain and management allowed Ford to pursue other interests.

Ford never fully left the farm throughout the early days of Ford Motor. Henry and his wife Clara spent much of 1905 and 1906 building up the old Ford homestead. They drilled water wells and planted trees. Henry started to see himself as a gentleman farmer. Henry had an aversion to farmwork and horses, so he hired hands at $1.50 a day to do the daily work. They expanded dairy production and raised hogs. Ford found time for some plant and crop experiments as well. Ford also loved to hunt deer and game birds. Clara and Henry were active birdwatchers on weekends. He and Clara found time for Detroit social events, and both loved to dance. In 1906, the family even took a cross-country vacation with stops at the Grand Canyon and Pasadena, California. Success also brought the first roots of philanthropy by Henry Ford. One of his earliest philanthropic checks would be $100 to the Tuskegee Institute. The concept of education, farm skill training, and industrial training was the perfect curriculum for a Ford-supported school.

During this period, Ford kept his organization focused on his new Model T development, cost reduction, and process automation. The Model T would be a true engineer-

6. The Model T and the Jesup Wagon

ing marvel with major advances in alloy steels, transmissions, metal castings, and starting systems. Henry knew where he was going but needed his army of engineers to turn the dream into reality. Ford was now freed for what he did best — creativity. He was a dreamer and an innovator more than a manager or an engineer. Like so many great business leaders, he knew how to pick managers who could get things done. His organization in 1907 had the industry's best engineers and managers. He also had the best suppliers. Suppliers like the Dodge Brothers and Fisher Body allowed Ford to develop and expand his role as assembler, allowing suppliers to handle complex products such as engines. Eventually this would allow Ford to master mass-production techniques of the assembly line.

Ford continued to pick suppliers with great care. He chose suppliers for the long haul and showed a true genius in supply chain management. The strength of his supplier chain is often overlooked in Ford's success. Standardized subassemblies allowed Ford to mass produce and would give rise to the modern assembly line. Most of his suppliers became good friends. His newest was Harvey Firestone and his tire company, which would be the sole supplier for the new Model T. Ford had met Harvey Firestone years before when Firestone was selling carriages in Chicago. Firestone had invented a pneumatic tire that could be easily changed, which Ford used on the Model N in 1905. For the Model T, Firestone supplied a new demountable rim for easy changing of tires. Firestone demonstrated the same genius of Ford, Edison, and Carver in inventing new products. Firestone would not only become one of Ford's trusted suppliers, but his best friend, sharing in Ford's "chemurgical movement." The Model T had some green "chemurgical" components as well, such as the use of Spanish moss from southern swamps to stuff the seats. Spanish moss was water-resistant, and it repelled insects, mildew, and bacteria. It grew around Tuskegee as well, and Carver had suggested it to mulch the clay-based soils.

One of the key steps in the development of the Model T was Ford's use of high-strength low-alloy vanadium steel. The story of this component of the Model T is part of the Ford mythology promoted later by Henry himself. Still, vanadium steel played a central role in the development of the Model T and the automobile industry. According to the old story, while observing a racing accident in 1905, Henry Ford was enticed by the good condition of a wrecked French car. He had the steel analyzed and discovered a high amount of vanadium in it. This vanadium steel offered a solution to Ford's design of a big, lightweight, low-cost car. Vanadium strengthened the steel, allowing for a thinner gauge (lighter weight), thus reducing vehicle weight with high strength. Interestingly, decades later with the energy crisis, the automobile industry would rediscover the high-strength low-alloy steel to reduce weight.

The first Model T prototype was too heavy for the size Ford wanted. He wanted a five-seat car, which required far too much weight and power if two-seat Model N technology was used. Even without the mythology, the real story behind Ford's persistence to reduce vehicle weight through better steels is amazing. Wills had been looking into alloy steels like nickel steel, which had been used in ship armor, but he could not significantly reduce the weight. Wills was even working with J. K. Smith, a former consulting British metallurgical engineer. Smith had become chief metallurgist for American Vanadium Company of Pittsburgh and had visited the Ford plant in 1906 to interest him in

The Green Vision of Henry Ford and George Washington Carver

A Farm Scene of a Family Visiting in an Early Model T Ford Touring Car. Oil, 16.5 × 16.6 in. A Norman Rockwell–commissioned painting of the Model T's relationship to the American farm (from the Collections of The Henry Ford, Benson Ford Research Center, Photograph THF95622).

vanadium steel.[5] Whatever the inspiration for vanadium steel, Ford's vision drove its development. Ford Motor became the pioneer in research on vanadium steel and had experimental batches made for testing in Detroit. Wills discovered that the use of vanadium steel with the proper heat treatment could cut the weight of a normal steel-designed car by 50 percent. The five-seat prototype had a total weight of 1,200 pounds with vanadium steel used in engine parts, axles, sides, and doors. This would allow Ford to replace the Model N with a bigger, stronger, cheaper, more fuel-efficient car.

The next step in late 1908 was the development of production batches of vanadium steel. Ford contracted to work with a small alloy steel company in Canton, Ohio. Henry Ford went to the melting of the first production-size lots of vanadium steel. The first production batches failed to meet Ford's specifications, and developmental work was needed in the processing. Ford's managers wanted him to hire a metallurgical engineer to work full time on this project. Ford, however, had a serious bias against university-educated engineers and scientists. Almost to prove his point, Ford assigned a former sweeper, John Wandersee, and told Harold Wills to "make a metallurgist" out of him. True to form, John Wandersee went on to have an outstanding career as a metallurgist and made vanadium steel a reality. With vanadium steel, Ford was ready to make his Model T a reality. Just as important was Ford's technical and financial support in the development of vanadium steel which changed American industry. Vanadium steel became a major advantage for American armor and cannon makers during World War I. Vanadium steel help solve problems with the Panama Canal when it was incorporated in the hinges of the locks.

Henry Ford and his toolmaker and now chief designer, Harold Wills, had been working on the new design that would be the Model T in their "experimental room" at the plant. The experimental room was limited to a handful of Ford's handpicked designers. Many of these men would change the nature of American industry and manufacturing. Most importantly, these designers would become the future greatness of Ford Motor Company. Ford had a blackboard installed for his directions. The experimental room was Henry's new workshop and his first stop every morning. This is where Ford spent most of his time versus his assembly plants. Ford brought his mother's rocking chair to the room for good luck. He demanded that wooden models be made of all parts as opposed to depending on blueprints. For the Victorian Ford, physical touch and feel were part of the creative process. He created in three dimensions and felt that two-dimensional blueprints were a roadblock to the creative process. Blueprints were fine after the product was developed. Once the car prototype was ready, it faced endless road testing from Henry Ford.

The Model T was revolutionary in every way. It revolutionized automotive engineering, steelmaking, car design, auto manufacturing, aviation, tank design, and the very culture of America. The Model T would be the first car with a steering wheel on the left side (since the standard was to drive on the right side of the road). It was the first full-block, four-cylinder twenty-horsepower engine. It could achieve forty-five miles an hour on the unpaved American roads. It had a new suspension system to absorb road shocks and an enclosed transmission system. Ford pushed hard to replace the open gears and hard shifting in all cars of the period. Stripped gears seemed to be part of the early automobile experience. Ford wanted a new planetary enclosed transmission, and he had a crew working on it, headed by a young draftsman, Joe Galamb. The planetary gearbox for the Model T was an engineering marvel. By World War II, it was adopted for tanks and aircraft.

It had a type of simplicity that allowed the average male of the time to make repairs. The $850 Model T with five seats made it a family car for the masses. Out of the experimental room came Ford's operating management. What soon became apparent to Ford and his operations manager, Charles Sorensen, was that the Model T would require a new organization to manufacture and sell it. Joe Galamb became his chief designer and John Wandersee the director of metallurgy. Henry Ford is often given credit for mass production and part standardization, but that is truly an oversimplification of history. Mass production and standardization go back to the 1600s and Arsenal of Venice.[6] In the 1800s, standardization had already led to mass production. Still, Henry Ford's automotive assembly line was every bit as revolutionary. Ford and many others in the Ford organization would later claim the concept of the assembly line as their own. Ford created his own mythology of adopting it after a visit to the automated Chicago slaughterhouses. In reality, the Ford assembly line was as much evolutionary as revolutionary, but it clearly came out of the vision of Henry Ford himself.

It was born out of Ford's obsession with daily production and costs as well as his vision of manufacturing a car for the masses. Ford's Model T forced weight and cost reduction. His success with vanadium in steel drove him to launch a major metallurgical research program, resulting in the use of molybdenum in steels to further reduce vehicle weight. Out of this obsession arose the new industry of alloy steels. Ford's obsession and vision, combined with a driven handpicked team, led to the Ford assembly line. The assembly line was not Ford's invention, but Ford made it pay dividends for the first time. The assembly line was only a means to an end. The team often chased a few cents' savings per vehicle. It represented efficient and economical mass production on a scale never seen before. For Ford, like Steve Jobs of Apple, obsession with a vision was the mother of invention. In many ways, George Washington Carver had the same obsession and vision. But most important, Ford created a lean organization.

Ford's vision and obsession translated throughout the organization. One of Ford's foremen, Charles Lewis at the Piquette plant, started to rearrange the order and supply of parts to speed up the production of the Model N. Charles Lewis was not alone; other foremen worked on increasing productivity and improving handling in their areas. One department designed a specialized drill press that could drill ten holes at a time. Combining operations by specialized machines became a standard for improvement. The simplification and reduction of materials handling further interested Ford's operations manager, Charles Sorensen, who daily reported production numbers to Ford. The obsession spread to Fred Diehl of the purchasing department, who in turn made on-time purchasing and delivery a science for the Ford plant. This is why many today (including Toyota) see the original Ford system as the real root of just-in-time manufacturing. The corporate obsession spread to Ford's sales manager, Norval Hawkins. Hawkins pushed the production department to meet his orders. He shortened the communications to the plant to ensure that production schedules reflected sales forecasting. This system would be the root of today's material requirements planning inventory systems. Furthermore, the evolutionary process of assembly-line development continued over the years. Another Ford foreman, William Klann, extended the system to parts in 1912. Klann's use of the assembly line to produce magnetos is considered the first true Ford assembly-line operation.

While he launched his Model T at the Piquette Avenue plant, Ford had been planning his new Highland Park Model T plant since 1906. In the later years of the Model N, Ford

6. The Model T and the Jesup Wagon

was having problems keeping up with consumer demand. He had started Model T production in 1908 in the Piquette plant with improved processing, but he could not meet demand or maximize efficiency in the Piquette plant. After rejecting a number of factory designs, Ford hired an industrial architect, Albert Kahn, to design a revolutionary factory worthy of the Model T. Kahn had become famous with his design of the Packard Motor factory. Again, as with the overhauled Ford system, Khan would turn Ford's vision of a lean factory into a reinforced concrete factory. Ford wanted, in particular, to design the layout to encompass all the efficiency and handling improvements of his operating management team. But even Highland Park would become evolutionary starting in 1910 and developing into full assembly operation by 1913.

The basic layout of Highland Park required wide aisles and high ceilings to allow cranes and conveyors to move materials. The plant had four stories. One of the reasons for using four floors was to use gravity-fed supply systems to automated lines. Ford laid out the factory along functional lines. He assured ease of supplying parts and handling materials at all points. Even the manufacturing procedures evolved with more Model T production. Ford started Model T production in manufacturing cells (a procedure many are moving back to today) in 1910 with teams of function-specific workers moving instead of the cars. Sales, however, were doubling every year, forcing a need for even faster production to meet demand. Ford moved more and more subassemblies to supply a moving assembly line to match demand. In 1912, Clarence Avery, a University of Michigan graduate and a teacher of Ford's son, Edsel, was brought into the organization. In eight months, he worked every job in the plant and was asked to time a full assembly operation for Model T production so a chain drive system could be added. In 1913, a fully chain-driven assembly line was in place.

The assembly line orchestrated the flow of 1,500 parts along a moving line into a finished Model T, cutting the car production time from twenty minutes to thirteen minutes. Skill levels allowed for untrained workers to be hired. By the end of 1913, the Highland Park plant employed 13,000, making it the world's largest factory. Ford required a huge influx of immigrant labor to man his assembly line, which would employ many of the poor black farmers leaving Alabama. The full application of the assembly line allowed Ford to drop the price of the Model T to $500. One out of every three cars sold in America was from the Highland Park assembly line. Ford became extremely wealthy and popular. Ford's income rose to over $7 million a year. The assembly line not only reduced the hard labor needed from his workforce, but made life easier for Ford.

Ford's life changed modestly from vacations at Niagara Falls to his first trip to Europe. When in London, he set up with the Shackleton Apiary to ship an array of European songbirds to his growing game preserve at Black Farms (later called Fair Lane). His interest in bird-watching expanded and he funded research into bird migration. He became a financial supporter of the Michigan Audubon Society. Ford became friends with environmentalist John Burroughs, who visited Black Farms in 1913. John Burroughs had been skeptical of industrial progress, but Ford won him over with his game preserve and gift of a Model T. Burroughs, Ford, and Thomas Edison teamed up to push the Weeks-McLean Migratory Bird Bill in 1913.

Unknown at the time to Ford, he had a fellow supporter of the Migratory Bird Bill, George Washington Carver in the South. Carver would also fight for the passage of the Weeks-McLean Bill. He had argued that insecticides were not as effective as birds in insect control. Carver argued that insecticides created more problems by killing off the bird population. In a 1914 bulletin, Carver noted, "Year by year the war on insects that threaten to destroy our farm, field, orchard, and garden crops, and often our personal comforts, becomes greater. We lay this condition at the door of some mysterious Providence. We do not seek the cause. If we did, every farmer and citizen would unite in one grand effort, not only to save, but to protect the birds, the greatest insect destroyers known."[7]

Ford also moved into social reform with his extra time and money. He viewed himself as a lovable baron of industry who wanted to save both community and environment with industrialization. Ford helped Detroit feed the poor and support community projects. He started farms and schools to help poor wayward children receive an education and learn agricultural work. Ford was showing the positive side of capitalism after decades of bad press generated by the robber barons. At the same time, Ford had started to think of a great outdoor museum for children to see farm machinery and science in action. He also started to purchase school desks and equipment from his grade school days. Ford believed that America's youth had been let down by the growing public system. He wanted a return to the basic education of his youth. He was already looking ahead to a new life mission.

7

The Industrialist and the Professor: Capitalism and Agriculture

"Business should be on the side of peace because peace is business's best asset."
— Henry Ford

"Every time I pick up a paper I think of what Mr. Sherman said war was, words fail to describe the horror and suffering."
— George Washington Carver

With the success of Ford and Carver came new phases in their lives. They both had achieved a level of success that allowed them to look past the competition and focus on the need for the initial climb. For both Carver and Ford, there were more professional peaks to conquer, but now they built on their previous success. By the late 1910s, Ford and Carver started to chase new passions and strengthen old ones. Both men started to look more to contributing to society as a whole. Ford would find that men more than machines could unleash the full power of the assembly line. By 1916, Ford had become a national figure and folk hero, while Carver was on the verge of becoming the same. Both men would be called by the White House to play a bigger role on the national scene. Ford would start on his career of philanthropy, environmentalism, and the education of American youth. For Carver, he would see that teaching was to be the foundation of his mission as much as research. Ford would fail with his peace initiative to keep America out of World War I. Carver would find new food sources to supply a nation in war.

Carver's experimental station farm had become a local legend. His use of composed fertilizer gave amazing results, as did his experimental and hybrid crops. Farmers often visited to awe at Carver's success. He would amaze them with twenty-plus-pound cabbages and seven-inch-diameter onions. He was producing over 250 bushels per acre with Irish and sweet potatoes. His fertilizer of swamp muck, leaves, kitchen scraps, and cuttings performed better, increasing profits to as much as $60 an acre. Cotton yields moved from a third of a bale per acre to a bale and quarter per acre. Government agents were also impressed and asked to be mailed all of Carver's bulletins and newsletters. Still, the government bureaucracy continued to reduce aid to Carver.

The reduction of funds to Carver's experimental station after the Smith-Lever Cooperative Extension Act of 1914 forced the resourcefulness that had, in the past, brought forth creativity from Carver. Carver always, after a short period of remorse, used setbacks as motivation. In 1915, Booker T. Washington, the founder of Tuskegee, died, leaving Carver without his greatest backer and source of outside funds. With all their personal issues between them, Washington and Carver had merged into a dynamic duo. Washington was the salesman and promoter for Carver's creativity. Carver would have achieved national fame eventually, but Washington got him there much faster. Former president Theodore Roosevelt came to Tuskegee for Washington's funeral and spent a day with Carver, encouraging him to continue his experimental work.

It was not only Carver and the experimental station of Tuskegee that were challenged, but the Black Belt farmers as well. By 1915, the boll weevil was devastating black farmers and their cotton plants. It was a crisis point with farmers, landlords, towns, and whole counties facing bankruptcy. Some blacks would move north to the auto factories of Henry Ford, but most were stuck with their bankrupt farms. A nearby county had all farmers unable to pay taxes. However, it was the crisis of the boll weevil that would lead to a solution and economic freedom. Carver had been unsuccessful at making major shifts in the attitude of the cotton farmers.

Carver the mystic had early on seen a real opportunity in the plague of the cotton boll weevil. He believed it was a plague sent from God to change the hearts of the farmers who resisted change. When the boll weevil hit nearby Coffee County, one farmer took up Carver's idea to switch to peanuts. The farmer produced 8,000 bushels of peanuts at one dollar a bushel and got out of debt, which caused a movement of others using those same peanuts for seed.[1] Years later, the town of Enterprise in Coffee County, in the heart of the Black Belt, erected a monument with the legend, "In profound appreciation of the boll weevil and what it has done. As a herald of prosperity, this monument was erected." Coffee County would become one of Alabama's wealthiest by 1920 with a new $25,000 peanut-shelling plant. Peanuts also offered the farmer flexibility for animal feed. Farmers now started to look at all of Carver's experimental work. The Black Belt of Alabama was finally changing, and Carver was now seen as the leader of that change. Carver saw a new opportunity to enforce the faith of these now economically improved farmers.

For Carver, the world was the Kingdom of God, and his laboratory was part of the kingdom. Henry Ford believed the same, but for Carver it took on a level of mysticism. The challenges of the boll weevil reinforced Carver's self-image as a type of economic Moses. His mission to save his people economically and spiritually came to be a new source of passion. Carver talked of this mission in coming to Tuskegee, but he saw it as a quiet role through his science. Like Moses, Carver had reluctantly accepted the more public role after the death of Booker T. Washington who had a fiery and public image. He clearly drew strength from this mystical view, but it would often draw scholarly criticism of his work. Any study of Carver cannot separate his work and his faith. Carver came from families of great faith, and his college professors were men of faith. The image of Moses and Carver's work was not lost on historians or on preachers. One of the most influential black preachers of the 1950s, C. L. Franklin, also saw a Moses in Carver: "In every crisis God raises up a Moses. His name maybe Moses or his name maybe Joshua or his name maybe David, or his name, you understand, maybe Abraham Lincoln or Fredrick Douglass or George Washington Carver, but in every crisis God raises up a Moses, especially where the destiny of his people is concerned."[2]

The idea of Carver the mystic is an exaggeration of more recent biographers. Carver was a Bible Belt Christian who saw God in all. Carver didn't see the world in mystical terms, but merely as the Kingdom of God. Carver was a simple Christian with a strong faith, more Benedictine monk than mystic. He started each day with a "long talk" with God. He believed in the Benedictine motto of "prayer and work," seeing them as interrelated. Interestingly, the focus on waste elimination and finding independence in nature's products was Benedictine. Carver saw work as prayer and often asked God for direction. Carver spoke in terms of God directing his scientific work. Talk of "God giving him the formula" or "Praying to God for a recipe" seems strange in today's language. He saw God as inspiration, not a talking voice, although he often used terms like "God said to me or God told me." Similarly, putting his career in terms of a mission from God seems strange to some today, but he had a simple God-based view of the world. He saw science in the Bible as well. He was a typical Bible Belt Christian, and like Henry Ford, he was not defined by any denominational doctrine. His Victorian religious thinking was out of vogue by the 1900s, which became a source of potential criticism for him as a true scientist. Still, even modernists must see him as a type of Moses to his people. The decades of devastation from the boll weevil required a leader. It brought his people to their knees looking for help.

The impact of the boll weevil was nothing short of complete desolation. By 1904, Texas cotton farmers had seen a 60 percent drop in yields as the boll weevil moved north. In 1910, Greene County, Georgia, produced 13,862 bales of cotton; in 1916 it fell to 11,864; in 1921 it fell to 1,487 bales; and in 1922, production was at a low of 333 bales. More southern counties had been destroyed by 1917.[3] The land could not support the population, creating the "Great Migration" of blacks in 1916 and 1917 to northern cities, particularly Detroit and the auto factories of Henry Ford. Things got worse in these years with flooding washing away what good topsoil was left. Carver had been warning about the progress of the boll weevil since 1906, but cotton remained king until farmers were faced with almost total loss. Eventually, from 1916 to 1928, more than 1,200,000 blacks left the South as King Cotton fell, with about 120,000 going to Detroit. Another push of blacks came in the 1920s as the government restricted European immigrants and southern blacks were needed in the auto industry. The percentage of blacks in Ford plants was well over 12 percent by 1930.

The boll weevil was another area that had brought Carver and Ford together on a missionary level. Clearly, major farm problems were a source of common interest throughout their lives. Interestingly, Henry Ford had been following the devastation of the cotton crops and had set up a research project at his Ford Motor laboratory to look into insecticide development.[4] Ford Motor was threatened by the loss of high-quality cotton fiber and the increase in price. High-quality cotton fiber was also critical to the manufacture of tires by his friend Harvey Firestone. Ford asked his branch managers throughout the world to collect information on the boll weevil and cotton growing problems. Ford Motor put together extensive research on insecticides. Ford even looked at the potential of going into the production of the insecticide calcium arsenate, which was effective but too expensive for the poor farmers of the South. He also looked into chloropierine which had been successfully used in Algeria. Carver, of course, was approaching the problem from a different perspective. Carver looked for a new weevil-resistant cotton hybrid or an alternative crop such as peanuts. The boll weevil would, in the end, make the "Peanut Man" world famous.

It would be in 1916 that Carver published his most famous research bulletin, *How to Grow the Peanut and 105 Ways of Preparing It for Human Consumption.* Carver offered a solution and made it happen, but the idea to plant peanuts in the South was not totally new. Carver had been experimenting with peanut plants back in his Iowa State days. He extended his peanut research when he came to Tuskegee, since Iowa's climate was not well suited for growing peanuts, which cannot tolerate frost and need a long growing season. The South offered a more favorable environment for the runner-type peanut that offered high yields over the garden-type varieties of Spanish or Virginia peanuts. The Spanish variety also offered the possibility of two crops a year. Being of the legume family, peanuts were soil enhancers that added nitrogen, which was a plus for the poor soil of the Cotton Belt. Peanuts had another advantage in the South in their ability to tolerate droughts. Yields could be as high as sixty bushels per acre.

The peanut offered an economic opportunity to poor farmers to integrate peanuts into their routine. Black farmers were laying cotton in the spring and summer, but the peanut could be harvested in late July after the cotton was cultivated. Carver preferred peanuts as a crop for this reason in his initial campaign to convert farmers. Carver had been promoting peanuts to farmers since the late 1890s, but they resisted change until the boll weevil reached them from Mexico. The peanut had been grown as a novelty crop in Virginia since colonial times, but it had found limited uses. Peanuts were actually native to South America but had made their way to Africa via the trade routes. Through the slave trade the peanut traveled from West Africa to North America. The African name for peanut was "nguba," which was the origin of the word "goober." They were not totally unknown in the Deep South where a few vines might be planted for snacks for the children who enjoyed eating them. Also in 1901, a roasted-peanut vending machine was introduced, and peanut candy entered the marketplace, creating a growing demand. Roasted peanuts became standard fare at America's ballparks. Peanuts also offered a source of vegetable oil, and the remaining pressed cake was an outstanding animal feed.

As a mycologist, Carver had been watching the northern progress of the boll weevil out of Mexico from 1902 when it crossed into Texas. Carver was looking at possible weevil-control methods with little success. He thought the better plan was to look at alternative plants and move out of cotton. He started to study sweet potatoes, soybeans, and peanuts. In 1903, Carver saw that the resistance to plant peanuts, soybeans, and sweet potatoes was market related. Sweet potatoes and peanuts offered the best chance of expanding food markets. He launched a research effort to expand the uses. He focused on recipes and cooking early on with the idea that struggling farmers could at least feed their families from the new crops. He involved the girls' cooking classes at Tuskegee in improving his recipes. The girls became famous for serving all-peanut-based meals to Booker T. Washington and important guests.

In 1916, the peanut (*Arachis hypogea*) was a substitute for cottonseed oil and rendered a by-product of a press cake that could be used for animal feed and sold at thirty-five dollars a ton. Peanut oil was a high-energy replacement for common vegetable oil, but it

7. The Industrialist and the Professor: Capitalism and Agriculture

also came at a higher cost. In the 1880s, Otto Diesel used it as a fuel in his new Diesel combustion engine. Peanut oil tended to be superior to most vegetable oils, but the harder peanuts made pressing difficult and expensive. Peanut oil with its high smoke point made it ideal for deep frying, which was popular in China. The same property made it popular in making popcorn. Peanut oil also was outstanding in the production of soap, but again it was higher in cost than the cheap imported tropical oils used in soap making. Carver even studied safflower, castor, and sesame oils for potential uses in soap making.

Carver's genius was rooted in his understanding of science and the advances being made to find new products for alternative crops. Linoleum had been a product which Carver improved on in the 1860s. It was a high end substitute for marble and had been used in wealthy homes as well as in luxury liners such as the *Titanic*. It was made from linseed oil, pine rosin, and cellulose flour heated to thicken it. Then pulverized wood was added, and the rubbery mass was rolled into a sheet and dried. Carver created a similar process using peanut oil, and instead of wood, he used peanut shells. As peanut food applications increased, linoleum offered a use for waste shells. Linoleum today is made from various vegetable oils, but peanut shells are still commonly used. Carver also pioneered the use of peanut shells as an intermediate abrasive for metal polished and as an insulating board component. Carver suggested peanut shell use in composting, which lowered odors. Today peanut shells are an ingredient in some pet litters.

Kellogg Company had started selling a high-protein food known as peanut butter in the late 1880s, which had been introduced at the 1893 Chicago World's Fair. Kellogg had even visited Tuskegee to talk to Carver about his work with the peanut. Peanut butter had become popular with manufacturers such as Kellogg and H. J. Heinz in the late 1890s. Peanut butter's high food value had made it popular with Arctic and Antarctic explorers of the time. Carver had started his research on peanuts in the 1890s while at Iowa State and was on the forefront of the uses of peanuts. Carver expanded food derivatives from the peanut to include coffee, milk, and various sauces. Sauces such as Worcestershire sauce were extremely popular in the time period.

Peanut milk was more perfect nutritionally than cow-processed milk. As early as 1915, Carver had fully developed peanut milk. Carver demonstrated the mix of protein and fat which could be adjusted to different products such as low-fat milk or buttermilk. It was easier to favor as well. It could remain stable in the heat of the South, which was important to poor farmers who lacked refrigeration. His work with peanut milk was exceptional, but it found market resistance in this country where milk was often in surplus. Peanut-based milk found applications in jungle areas where cows could not be raised. It also offered milk for poor black farmers in the South where storage was a problem because of the climate. Another exception was Henry Ford who actually preferred soy- and peanut-based milk to that of cows. Ford said in 1921, "The cow is the crudest machine in the world. Our laboratories have already demonstrated that cow's milk can be done away with and the concentration of elements of milk can be manufactured into scientific food by machines far cleaner than cows."[5] Both Carver and Ford spent their lives trying to replace cow's milk.

Carver improved on the production of peanut flour, which like that of Henry Ford's soybean flour, had a low carbohydrate level, making it good for a diabetic. Carver had sent some of his peanut recipes in 1911 to John Harvey Kellogg, the cereal king, which started a letter-writing friendship between the two. Kellogg was a strict vegetarian who believed the peanut to be a perfect substitute for meat. Kellogg was particularly interested in Carver's work to produce peanut milk. They exchanged letters between 1919 and 1926 on a range of topics. Besides peanut milk and flour, Kellogg was interested in sweet potato flour. He had worked with the government during World War I on Carver's suggestion for sweet potato flour. Personally, Kellogg was a vegetarian who often ate sweet potatoes. He would popularize a sweet potato recipe using a stuffing of corn flakes and marshmallow. Kellogg discussed diet with Carver, who suggested alfalfa salad as a dish for Kellogg's Battle Creek health sanitarium. Carver had hoped to convince local farmers of the table use of both peanuts and alfalfa, which would also supply nitrogen to the soil.

Peanuts would be a difficult sale in the short run, but Carver was gaining a reputation as the "Sweet Potato Man" by the early 1900s. Sweet potatoes offered a fairly ready food for the table. In 1918, carver summarized his work with a bulletin: *How to Make Sweet Potato Flour, Starch, Sugar, Bread, and Mock Cocoanut.* He had been developing the uses of sweet potatoes for almost twenty years. The sweet potato offered cash sales and table food. The sweet potato, like the peanut, originated from South America. Carver believed the sweet potato to be the best alternative crop for cotton farmers. The potato had a short time to harvest, which could allow two harvests in the Black Belt. The sweet potato grew well with compost, but fertilizer could multiply yields. Carver had done experiments with sweet potatoes as early as 1898. One of his first bulletins was *Experiments with Sweet Potatoes*. In this bulletin, Carver reported the results over two test plots. One was grown without fertilizer and the other with a phosphate/potash mixture. The unfertilized plot yielded 25 cents per bushel with a yield of 40 bushels. The fertilized plot yielded 50 cents a bushel with an amazing yield of 266 bushels.[6] Carver proved that sweet potatoes also made excellent animal feed, convincing him that it was the ideal crop for Cotton Belt farmers. Carver wrote more on sweet potatoes than he did on peanuts:

1898 (Bulletin 2)—*Experiments with Sweet Potatoes*
1906 (Bulletin 10)—*Saving the Sweet Potato Crop*
1910 (Bulletin 17)—*Possibilities of Sweet Potatoes*
1915 (Bulletin 20)—*Possibilities of Sweet Potatoes*
1918 (Bulletin 37)—*How to make Sweet Potato Flour*
1921 (Bulletin 38)—*Preserving Sweet Potatoes*

Early on, Carver had over a hundred acres of sweet potatoes. While sweet potatoes had a developed market, Carver needed to expand it to ensure good prices. He developed methods to produce sweet potato flour. He mixed this sweet potato flour with two-thirds regular flour in the Tuskegee dining hall. He was able to save the dining hall twelve dollars a day. Still the color of the flour was problematic to the average consumer. During the shortage of wheat flour during World War I, Carver was brought in as a consultant to help the military substitute sweet potato flour. He also suggested meat substitutes during the war. Carver continued to develop recipes for his sweet potatoes. The two most popular

7. The Industrialist and the Professor: Capitalism and Agriculture

are still used today—sweet potato pie and sweet potato doughnuts. Carver had found other uses such as library paste, dyes, instant coffee and others.

The First World War created shortages in dyes, rubber, and foodstuffs. The sweet potato offered a whole array of potential industrial uses. Carver would start into a new field of chemurgy, which used farm products to produce industrial products. Clearly, Carver deserves the title of first chemurgist. He issued Bulletin 33 during the war to help southern farmers become independent. He suggested switching out of cotton because of the coming of the boll weevil. In particular, he suggested sweet potatoes, peanuts, corn, soybeans, and cowpeas. The few farmers that switched in 1917 were rewarded handsomely. He also argued for composting to save hundreds of dollars on fertilizer.

In 1917, he was studying the making of rubber from sweet potatoes. The search for a domestic source for rubber would be taken up years later by industrial friends, Henry Ford, Thomas Edison, and Harvey Firestone. While he never fully unlocked the secrets of domestic rubber sources, he did contribute to the development of dyes, which had been a German monopoly prior to 1918. During the war, Carver was featured in national newspapers as his uses of sweet potatoes spread. He was given more freedom from teaching at Tuskegee so he could focus on research. Carver was now spending more time off campus giving talks and consulting. Congress had even introduced a bill to propagandize the use of sweet potatoes and sweet potato recipes to help reduce public use of strategic foodstuffs and materials. The $250,000 appropriation would die in committee as the war came to an end.

Interestingly, as Carver was looking for new recipes for peanuts and sweet potatoes, Henry Ford was switching his own diet to a more vegetable-based one. Actually both men shared the same beliefs in a diet high in fruits and vegetables and low in meat and dairy. To celebrate his new love of vegetables, Henry Ford planned a large banquet at a Detroit hotel to serve a meal made from carrots in 1912. The twelve-course meal consisted of carrot soup, carrot loaf, carrot au gratin, carrot torte, and carrot juice. Henry dressed up as "King Carrota." There seems to be little doubt that Carver and Ford were predestined to be friends. Later in their lives, Ford loved to eat meat and dairy substitutes from soybeans, and Carver from peanut by-products.

Money allowed Ford to pursue his passions and his interests to a much deeper degree than Carver. One of these was his estate at Fair Lane, which was to be a green-based home and nature preserve for bird-watching. The mansion had fifty-six rooms, eight fireplaces, an indoor heated pool, a stable for riding horses, a golf course, and a bowling alley. On the River Rouge, Henry Ford built a water-powered energy plant for the house. Thomas Edison dedicated the waterpower plant in 1914. Ford would become a promoter of waterpower versus coal-fired plants. The power plant at Fair Lane would be a prototype for a series of over thirty water-powered factories he would build in rural Michigan. The powerhouse also had a home laboratory for Henry to tinker in, a twelve-car garage next door, and a dock with an electric boat. His friend and naturalist, John Burroughs, helped him design the estate to be a world-class bird sanctuary. Today, Fair Lane is preserved on the Dearborn campus of the University of Michigan. By 1914, Henry Ford was moving on from industrialist to reformer, but there were still challenges for his business.

Henry Ford's first challenge came in 1914, as his mass-production technique that was doubling production yearly seemed to stall out. Actual labor productivity had not increased on the same scale as the number of cars produced. Ford had finally reached the point where machines could take him no further. The weakest link in the assembly became the men who manned it. The speed and autonomy of the line had defeated Ford's older workers. Absenteeism had reached more than 10 percent, requiring thousands of extra men to keep things running on a daily basis. Even more amazing was that turnover had reached more than 300 percent, requiring endless training and losses to the learning curve. While Ford was one of the best-paying employers, some were saying Ford was not sharing the wealth. Ford's response would make history. He doubled wages to five dollars a day and cut the day from nine hours to eight hours. Ford hired a labor expert who pointed out Ford's failure to address the human factor in his mass-production techniques.

The wage of five dollars a day made the auto assembly workers the highest-paid workers in the world. Steelworkers made $2 a day, glassworkers made $1.70 a day, and coal miners made about $2.50 a day. The result at Ford plants was a dramatic drop in absenteeism and turnover as well as impetus to the black migration from the cotton fields and south. The doors of Ford offered no color barriers, which were common in American industry of the time. The five-dollar wage had a revolutionary effect on American society beyond the great migration. The autoworkers were now able to afford the autos they were manufacturing. In addition, Ford went on to give consumers a "profit-sharing" reward of fifty dollars on every purchase of a Model T. This also took the steam out of the communist and socialist movement in the United States. Ford made capitalism a positive movement again in the United States after decades of trust busting and corruption made almost daily headlines. Ford furthered his image as an American folk hero, and Ford Motor gained more from it than any massive advertising campaign could achieve.

In fairness, the five-dollar-a-day wage came with some of Henry's own bit of social engineering. Workers were to be married or provide support for a family. Single men more than twenty-two years of age were to be of "proven thrifty habits." Ford also required a proof of sobriety and clean living. To enforce or at least guide his employees, he established the "sociological department." The sociological department started with employee publications on types of clean living and personal habits expected from being a Ford employee. Ford carried his social initiatives to extremes with fifty or more department investigators visiting workers' homes to photograph and record their living environment. Ford gathered support from Detroit religious leaders, labor leaders, and progressives. He added free medical and legal aid to help his employees. For immigrants, he had free education to learn reading and writing and pass the citizenship test. Ford brought in Episcopalian preacher Samuel Marquis to head up a type of corporate spirituality movement. Ford had plenty of critics of the strict Victorian social program, but the time was right for such a program. Many believed he was changing society for the better. Ford led the fight against liquor and smoking, which gained him many allies with the progressives of the time.

Another dark side of the sociological department was his anti–Jewish and anti-Catholic views. Some of this bias may well go back to his Irish Protestant roots. The sociological department tried to impose a set of values as a requirement to work at Ford Motor. Ford's opposition to smoking and drinking might not seem so oppressive today. Family background checks might not cause dismissal but would limit careers. Ford became a Freemason, although he used money donations to advance versus attending actual meetings. He also gave preference to hiring Freemasons. The Freemasons became a powerful group within management, and Catholic managers were often very restricted in their opportunities to advance. This dark side was indeed unfortunate for a man who would be a pioneer for blacks, the handicapped, and women in industry. Probably on balance, Ford did far more good for society than any harm.

Ford even went further in trying to bring his employees closer to the earth and farm. Ford saw gardening as part of the quest for self-reliance and his holistic view of a person. He made 35-by-60-foot plots available for all employees to garden in 1918. He further encouraged home-owning employees to have vegetable and flower gardens. He hired a company gardener to give gardening advice to his employees. He distributed bulletins, similar to Carver's, telling employees of the benefits of vegetables and how to plant and fertilize the. He launched a company program called "One Foot in Industry and One Foot in the Soil." Henry continued his personal vegetable garden at Fair Lane, while Clara built up major flower and rose gardens. Henry Ford would, like George Washington Carver, promote gardening as a necessary part of any school curriculum. It was reported that by the end of the 1930s, there were over 3,000 employee plots in use, and another 55,000 of his employees had home gardens.[7]

Ford's focus on immigrant education and citizenship was particularly popular. The effort combined Henry Ford's passion for Americanization of immigrants and his belief in a moral-based education. As thousands of immigrants poured into Ford's factories, he established the Ford English School. The school employed over 150 teachers in a unique educational program for immigrant Ford employees. The effort was modeled after an early effort by George Westinghouse in his Pittsburgh factory to educate immigrant workers. Ford brought in Peter Roberts of the YMCA to head up this school in 1914. By 1915, the Ford English School had graduated more than 5,000 students. The school methods were based on Ford's own love of the McGuffey approach to education. English lessons were couched with American history, moral values, home economics, and personal hygiene. Classes were coordinated with the work schedule to the ease of the employees.

Success changed Henry Ford personally as well. Ford now became part philosopher and national leader. In 1914, he donated a new hospital to the City of Detroit, but his real change came in his self-image as a social reformer. In 1915, Henry Ford would take on a personal diplomatic mission to Europe to hold the peace. While many have characterized Ford as a pacifist, he was probably more of an American isolationist. He broke with fellow Republicans on the need to prepare for war, which he believed would lead to war. Early in 1915, the *Ford Times* reprinted George Washington's farewell address with Henry's warning about foreign entanglements. By summer, Ford's antiwar quotes were in all the major news outlets.

Ford seemed to put his peace initiative ahead of business, as his executives begged him to stop using the company name in his efforts. Ford seemed to move from isolationist to pacifist as the nation and public opinion moved more toward war. Ford teamed up with an unlikely comrade in Jewish activist Rosika Schwimmer to rent a "peace ship" to Europe in late 1915. The ship was stocked with supplies, over fifty peace delegates, and forty-four reporters to make the crossing and hold educational lectures in Europe. Ford failed to enlist the full support of many, such as President Woodrow Wilson, his wife Clara, and friends like Thomas Edison, John Burroughs, and Harvey Firestone. The cruise was a disaster as Ford soon grew tired of the self-righteous academic types that surrounded Rosika Schwimmer. Henry blamed Schwimmer, and many believe she was the root of his anti–Semitic views.[8] Henry could often generalize groups by bad experiences with individuals. Ford was asked by President Wilson to run for the Senate, but that too ended in defeat.

Henry returned to Detroit, summing up the peace trip, "I didn't get much peace, but I learned that Russia is going to be a great market for tractors."[9] Returning to Detroit, he would set up a new company to build a tractor prototype (the Fordson Tractor). This was a project more consistent with his farming roots and life goals. Henry had been interested in tractors even prior to his car, having worked for Westinghouse Company selling and fixing steam-powered farm equipment. In 1910, Henry had even started designing and applying for tractor patents, but the automotive factory of Highland Park forced most of his attention. For years, however, Henry Ford would show up at tractor pulls and farm equipment shows as a pastime. Company stockholders had blocked Ford from entering the tractor business, but he worked on buying out the opposition over the next ten years.

In 1915, Ford got serious about the manufacture of a mass-produced gasoline tractor. He set a simple goal of a $200 tractor that could do the work of six horses. It would be the Model T of farming. Ford dreamed of a strong, powerful, lightweight, and affordable tractor. The best-selling Bull Tractor Company was priced at $645. The Ford design would have the simplicity and flexibility of the Model T. His best manager and metallurgist, Bill Sorenson, developed a new chrome steel to strengthen Ford's smaller design. In particular, Ford's work with stronger gears required by tractors opened up the new field of alloy steels, creating a new industry. Sorenson worked on this new chrome steel with a young metallurgist at Central Steel, Ben Fairless, who later became chairman of the board for United States Steel. The tractor design was based on the field problems that Henry knew well. In addition, his tractor would have attachments to grind feed, bale hay, husk corn, and be a power source for other equipment.

His gasoline-powered tractor would offer the farmer the same advantages as the auto consumer — reasonable price, durability, lightweight, and reliability. These were major advantages over the steam tractors that Henry had sold as a young man and would multiply his fortunes. In 1918, Ford had produced his first tractors, giving the first two to his friends, Thomas Edison and John Burroughs. By the end of the war, he had produced over 95,000 tractors and would build plants in Europe. By 1920, Ford could produce 400 tractors a day at his River Rouge plant. His tractor success would lead him into the new field of farm chemurgy. Ford estimated that it took thirty acres of grain to feed the six horses a year, which now were freed by the tractor for other uses. Ford now turned to exploring new uses for farm products.

7. The Industrialist and the Professor: Capitalism and Agriculture

Ford's success with tractors would inspire the northern root of chemurgy among farmers, as Carver's work in the South had started this new science. Now Ford joined other chemurgists in the search for new farm-based products. In 1916, Ford formed his own company to explore the use of alcohol from plants as a fuel for tractors, making the farm self-sufficient. This side company and his River Rouge Laboratory started decades of research in alternative fuels, natural plastics, plant-based chemicals, and food production. Henry hired his old friend, Dr. Edsel Ruddiman, from his position as dean of the pharmacy school at Vanderbilt University in 1926. Ruddiman had been doing research on legumes and soybeans, which he brought to the new Ford laboratory at the River Rouge plant. Ruddiman first produced a soybean vitamin biscuit for Henry. The simple lab would develop for Ford's chemurgical effort an array of new uses for farm products. Ford continued his first soybean lab at Greenfield Village using trade school students to man it.

This chemurgical movement captured Ford's attention in the years prior to World War I. Ford had developed another bias against the new breed of Texas millionaire oilmen and their bankers; and that, coupled with a new focus on the American farm, gave Ford yet another passionate project of alternative fuels to replace oil. Ford saw the oilmen and bankers as the natural enemies of the farmer. For Ford, the perfect inspiration was a mixture of resentment, prejudice, and vision. Ford believed that America would use up its oil due to the rate that cars were taking to the roads, but many laughed at his predictions. He predicted in 1916 that prices of gasoline would rise "to the point where it will be too expensive to burn as motor fuel. The day is not far distant when, for every one of those barrels of gasoline, a barrel of alcohol must be substituted."[10] Besides his hatred of oilmen, Ford believed that exhaust fumes presented health problems. Furthermore, Ford saw a synergy in the idea of alcohol-driven tractors harvesting corn for the production of clean-burning alcohol. It would be the first link in an agricultural assembly line for the nation. Ford formed a separate company, Henry Ford & Son, to explore alcohol production in 1916.

Ford had led the prohibition crusade in the state of Michigan, which outlawed alcohol production in 1916, causing the loss of many jobs. Ford hoped to turn those breweries to ethanol production for cars and tractors. He formed a research group to look into the development of alcohol fuel. His work was extremely visionary and is being revisited today. Ford took several trips to Cuba to purchase sugarcane plantations for alcohol production. He and Thomas Edison experimented with growing sugarcane in Florida for the same purpose. In the forests of Northern Michigan, he experimented with wood chips to make alcohol. One side innovation from that effort was the development of carbon barbecue briquettes that are still used today. With his Brazilian rubber plantation, Ford made wooden car parts from the cleared jungle wood.

While Carver was testing sweet potatoes for alcohol production, Ford was looking at a sweet potato variant from Germany for alcohol. The Germans, like Carver, realized that the higher sugar level in sweet potatoes made it the ideal mash for alcohol production. Ford's own success in getting the government to ban liquor production would come back to haunt him. Prohibition and the government would prove to be the end of his alternative and alcohol fuel program. The government was inflexible in its ability to manage different alcohol products. The Internal Revenue Service would, in the end, prohibit Ford from distilling alcohol for any purpose. Ford, however, would renew his interest in alcohol-based fuel in the 1930s.

Inspired by his traveling "vagabond" friend, Harvey Firestone, Ford launched a program to find an alternative plant for rubber production. This search for an alternative had started in 1916 during a vagabond camping expedition of Ford, Firestone, and Edison. Edison would take on the research challenge. It was in 1916 that Edison brought George Washington Carver to Menlo Park to discuss making rubber from sweet potatoes.[11] Edison would inspire Carver to look at goldenrod as a potential source of rubber latex. Firestone had to double the price of his tires after World War I because of the British monopoly on crude rubber. Ford enlisted Thomas Edison's help and put his Rouge Lab on the project as well in 1923. Edison, Ford, and Firestone set up additional labs in Fort Myers near their summer homes and plantations. In addition, Ford built a lab at Ways Station, Georgia, in 1924. The work progressed slowly to find alternative rubber sources. In 1927, Ford purchased a rubber plantation in Brazil to experiment with more efficient plantation growing and management. Ford named his Brazilian plantation Fordlandia.

As Ford moved into what seemed an endless array of projects, Ford Motor roared on by doubling sales in 1917 alone. Tractor sales soared as Henry Ford became the nation's first billionaire. His success meant success for his favorite supplier, Firestone Tire and Rubber. Edison had already become Ford's best friend. In 1914, Henry Ford, Thomas Edison, Harvey Firestone, and John Burroughs vacationed together in Florida at Edison's Fort Myers home. Eventually, Ford and Firestone would build winter homes there. Ford and Edison grew closer over the years, with Ford financing many of Edison's projects, as Edison's fortune had been depleted. Ford also gave Edison a new Model T every Christmas; and in later years, he gave Model Ts to all in the Edison family.

The four friends spent time exploring and camping in the Everglades that year. The trip would be followed by years of camping trips around the nation by the "Four Vagabonds." The longer trips included the West Coast, middle Atlantic states, and New England. They often added famous guests such as presidents Calvin Coolidge, Herbert Hoover, and Warren Harding, or stopped to visit prominent men such as Luther Burbank in California. At a 1921 campsite around Hagerstown, Maryland, President Harding joined them for a dinner of lamb chops, ham, corn, potatoes, and biscuits. John Burroughs called Ford's food wagon the "Waldorf-Astoria on wheels." After dinner, President Harding went to Ford's tent for a nap. When President Harding died of a heart attack in 1923, Firestone, Edison, and Ford took a camping trip to the funeral in Marion, Ohio. The group then went to Edison's birth home at Milan, Ohio, and on to Ford's Dearborn, Michigan, estate before traveling to Northern Michigan.

Future president Colonel Dwight David Eisenhower joined them at the campfire in 1919 to discuss a national road system for the military. Press photographs included fireside chats and chopping wood. The families of the vagabonds would remain lifelong friends with Bill Ford, the grandson of Henry, who married Martha Firestone, the granddaughter of Harvey, in 1947.

7. The Industrialist and the Professor: Capitalism and Agriculture

The trips were hardly rustic excursions into the wild. These caravans often had up to fifteen cars with reporters following. Interestingly, not all were Fords. Edison often had a Cadillac and Firestone a Pierce-Arrow. The lead cars had the four vagabonds and their chauffeurs. In later years, they included the wives and family members. One Model T was equipped with a mobile kitchen and a refrigerator to store fresh steaks, eggs, and fruit. John Burroughs, in particular, loved a good steak. Ford often brought along a Japanese cook to prepare meals. The dining table could seat twenty. Edison, on the other hand, tended to be a light eater; and like Ford went through different dietary phases. A power generator was included to provide electric lighting. There were large monogrammed tents for each of the vagabonds. They did take back roads by which Edison and Ford loved to explore streams, rivers, and the associated water mills. These trips offered time to dream and find inspiration for new projects. They also helped to make the family automobile vacation part of the American psyche.

There were many other trips such as Ford and Edison's trip to the 1915 Panama Pacific Exposition in San Francisco. The men and their wives took Ford's private railroad car to California and then motored the coast to the exposition on "Edison Day" at the fair. Harvey Firestone would meet them there. The three families would take part in the "all-electric kitchen" dinner at the exposition. The three families spent ten days at the Inside Inn of San Francisco followed by a trip to Santa Rosa to visit the famous botanist, Luther Burbank. This was the trip that would turn Edison's interest to that of botanic research.

Burbank was a Victorian scientist like George Washington Carver, but he lacked the respect of many university-trained scientists. Carver had modeled his graduate research after Burbank's hybrid work. Interestingly, journalists often called Carver the "Negro Burbank." Both Burbank and Carver would suffer in the 1930s from the weak opinion of Franklin Roosevelt's secretary of agriculture that neither man did noteworthy research because of his hatred of the chemurgical movement.[12] Burbank and Carver, however, had earned the deep respect of Henry Ford who saw them as genius worthy of Ford's personal pantheon. Ford had offered Burbank a position in his new rubber research, but Burbank was happy in California. No doubt Carver's name was discussed, and Edison did indeed follow up with an offer to Carver to come and do rubber research at Menlo Park, New Jersey.

This day with Burbank in 1915 would cover two important topics for the men that would be themes for the rest of their lives. First, they discussed finding a North American alternative plant for rubber production. These industrialists feared the cutoff of latex raw rubber from South America and Asia as the war spread worldwide. The other topic was the manufacture of alcohol to power cars in America. For many, this meeting would take Edison into the world of botany and be the start of the industrial branch of chemurgy. Ford was pleased to befriend Luther and a year later would send him one of the first Fordson tractors produced. Later, Burbank would sign the cornerstone of Ford's museum, and Burbank's office and birthplace would be transported to Greenfield Village. Ford's gardens at Fair Lane would reflect much of Burbank's hybrid technology. Henry became a collector of Burbank's plates and drawings.

One of the most important projects that evolved out of the vagabonds' visit with Burbank was a new emphasis by Ford, Edison, and Firestone on economic botany, which ultimately evolved into the chemurgical movement of the 1930s. Harvey Firestone feared a rubber shortage as a result of the upcoming U.S. involvement in the war in Europe in 1917. Economic botany was not new to Edison. He already had sent engineers around the world for bamboo and other plant fiber to make carbon filaments for his incandescent lightbulb. Edison may have even been ahead of Ford and Firestone in his interest in botany, having visited Burbank two other times in 1915.[13] Ford would shortly start his study of soybeans and hemp for industrial uses. The trips of the vagabonds slowed after the death of John Burroughs in 1921, but it wasn't until 1924 that Ford, Edison, and Firestone made their last trip.

The vagabonds were really a group of "green" friends with campfire discussions on ecology, conservation, chemurgy, farming, and nature. Firestone and Ford had deep roots in farming. When asked what made for greatness, Firestone stated merit, hard work, service, leadership, and farm life. For him, "the American farm represents the spirit and fundamentals that have made America so great a nation."[14] The pneumatic tractor tire remained a personal project of Harvey Firestone's his entire life as did his search for American rubber. After Firestone's death, his farmhouse was maintained by the visitor's center for its Columbiana, Ohio, test track. The Firestone sons moved the farmhouse to Henry Ford's Greenfield Village in 1985. Appropriately, the Firestone, Ford, and Edison farms would be united at Greenfield Village.

One theme of these trips was Ford's interest in hydroelectric power throughout the nation. Hydroelectric power was often the topic at the campfires between Ford and Edison. They had often said there was enough running water wasted in the world to supply the power needs of all its countries. Likewise, environmentalist John Burroughs was very supportive of this clean energy alternative to coal-fired power plants. In 1919, Ford, Edison, Firestone, and John Burroughs camped on Green Island, New York. It was a favorite fishing spot of Edison's near his Schenectady General Electric plant. Ford had decided to build a radiator plant there using a new hydroelectric power dam at the site. The government had built the dam earlier and Ford would pay rent. He planned to use the old Erie Canal as part of the transportation path to Detroit. The vagabonds carved their initials in a stone that would become the future cornerstone of the factory. This radiator plant of about 1,000 employees would become a model for Ford's water-powered village industries in Michigan. Ford and Edison returned with President Warren Harding to dedicate the opening of the plant. Green Island was a statement about Ford's support for hydroelectric more than a decentralized village industry, since the distance from Detroit added costs. Edison had also talked of building a Green Island manufacturing plant, but he had lost General Electric to New York bankers and was short on cash. Ford also had a branch assembly plant operating in St. Paul, Minnesota, with hydroelectric power. Here Ford was also renting the use of a government dam.

Most of the financial support of Edison's later life research came from Ford. Edison had never fully recovered from his loss of his company, General Electric, to New York banker, J. P. Morgan, in 1890s. Ford had financed an early developmental company to

7. The Industrialist and the Professor: Capitalism and Agriculture

produce hard rubber batteries for Model T's. Rubber battery cases had grown into a major use of rubber after tires. Ford had also financed Edison to further research battery improvements, hoping to bring an all-electric car onto the market in 1914. Ford had always been conflicted about the pollution his factories and cars created. He particularly hated the fumes of his gasoline engines and wanted to use clean-burning alcohol. Ford also hoped he could resurrect the clean electric vehicle his Model T had knocked out of the market, but he needed an improved battery. Actually in 1895, Thomas Edison had developed a three-wheeled electric car, but he never fully perfected the battery power needed.

For Christmas 1913, Ford gave Edison a Detroit Electric car. A few months earlier, Ford purchased an electric for Clara. Ford loved the clean electrics, but he tired of having to tow Clara's electric car when she ran down the charge on her trips from Detroit to the Dearborn farm. Henry eventually installed a battery recharging station in Dearborn. In January of 1913, Henry, Clara, and Edsel visited Edison at his West Orange, New Jersey, research center. Henry and Edsel used the extensive library there to research the possibility of producing an electric auto. They came to believe they could produce an electric car that could go 100 miles at twenty-five miles an hour on a single charge. The project was to be Edsel's first time to lead as an executive of Ford Motor.

The electric car project was a joint one between Ford Motor and Edison Storage Battery Company. Edison was able to develop a major advance in battery design, using iron-nickel versus the lead-acid batteries. In 1914, a prototype Ford-Edison electric car was produced. The Edison battery had the power but it was heavy, which hurt vehicle performance. The Ford-Edison electric auto, using a Model T frame, weighed 1,350 pounds and had a top speed of a mere seven miles per hour. A second car was produced but still was short on the performance needed to help it compete with the Model T. The coming war and rubber shortages seemed to change the interests of Ford and Edison into more pressing economic matters such as a domestic rubber source.

With the early encouragement from fellow vagabond Harvey Firestone, Edison started some early rubber research in 1917 because he was a major user of hard rubber for his batteries. The effort slowed during America's entrance into the war as other efforts once again took priority. Surprisingly, it was after the war in 1922 that the rubber shortage came again to the forefront. The British rubber cartel announced a plan known as the Stevenson Plan to reduce exports and raise prices. The move forced Secretary of Commerce Herbert Hoover to bring together Harvey Firestone, Thomas Edison, and Ford to craft a strategy. They formulated an overall strategy, with fellow vagabond and president of the United States Warren Harding, to create a rubber research bill. The bill supplied $400,000 for rubber plant research. Ford and Firestone would pressure Edison to build a rubber laboratory at their summer compound in Florida, and Edison would look to bring men like George Washington Carver and Luther Burbank into his organization.

Edison showed some initial success with milkweed, and Ford quickly put his Dearborn team on it. Edison was also interested in guayule, which was native to northern Mexico. Guayule would require the heat of Florida and large amounts of land. Edison became very high on the potential for guayule latex. Ford began purchasing land in Hendry County to try commercial farming of guayule. Ford and Edison loved the press,

Henry Ford's 1914 electric car prototype using a Thomas Edison battery (from the Collections of The Henry Ford, Benson Ford Research Center, Photograph THF95629).

and their moves created a land boom in southern Florida. However, the effort would eventually fail. Rubber research would consume Edison until his death in 1931. Henry Ford promoted the research by building a Fort Myers lab for Edison. Ford believed that the project would energize an aging Edison. Mrs. Edison noted in the late 1920s, "Everything has turned to rubber in our family. We talk rubber, think rubber, dream rubber. Mr. Edison refuses to let us do anything else."[15] Edison tested over 3,000 plants, but found his best success with goldenrod. He sent the rubber to Firestone who made a set of tires for Edison to put on his Ford. Henry Ford started to build an extensive new rubber laboratory for Edison in Ways, Georgia, but Edison would die before completion. Years later, Ford would hire Carver as a consulting scientist at Ways.

8

Massive Assembly, Peanuts, and Aircraft

> "No individual has any right to come into the world and go out of it without leaving behind him distinct and legitimate reasons for having passed through it."
>
> — George Washington Carver

In the 1920s, both Ford and Carver would reach a peak in national prominence. Ford would build his ultimate application of the assembly process at the River Rouge factory. River Rouge would be America's citadel of industrial might, the assembly line and vertical integration (owning the raw material supply chain) on an epic scale. By 1926, it would have 75,000 employees producing 4,000 cars a day from raw materials and would become the world's largest factory. Ford took glass- and steelmaking to new efficiencies through integration. For example, Ford produced window shield glass at twenty cents a square foot, versus thirty cents to $1.50 a square foot if purchased outside.[1] Ford's mission to create an industrial and historical pantheon of American accomplishment would overshadow even this manufacturing achievement. For George Washington Carver, he would burst on the national scene as "mister peanut." He would make two outstanding presentations — a 1920 one at the peanut industry convention and his 1921 speech in the U.S. House of Representatives. While these speeches brought fame to Carver, his foray into politics would increase the criticism of his work. Ford had already learned about the double-edge sword of political fame.

With his personal reserve and humbleness, Carver would seem unlikely to be pulled into politics, but his personal mission to improve the life of his people, like Moses, necessitated it. The problems of this political foray and the background of it have been overlooked by his biographers. For Carver, it was a matter of freedom from economic slavery for his people, and it was economics that took him to Congress to testify. Tariffs were the background of American policy. The political division on tariffs was second only to slavery in its divisiveness. Many believe it was the economic cause of the Civil War, since the slavery and plantation system fell into the same political side as the anti-tariff group. To enter this charged political battlefield was to open oneself to the harshest political attacks. Carver would go to Congress to support peanut tariffs in the manner of Abraham Lincoln.

Abraham Lincoln is most often remembered for freeing the slaves, but many consider his contributions to American economic growth on an equal par. Lincoln, a former Whig, actually ran on a strong economic policy of American protectionism through scientific tariffs. Lincoln had been the first national candidate since Henry Clay in the 1820s to have the united support of labor and manufacturers under the "American System." Lincoln's disciples in Congress such as congressmen "Pig Iron" Kelley, James Garfield, William McKinley, and Thaddeus Stevens, carried the Republican protectionist banner in the time period between Henry Clay and Herbert Hoover. These sixty years of Republican tariff policy protected, promoted, and grew segments of America industry and agriculture deemed critical to American economic growth. These men also demanded congressional oversight to ensure that profits gained from tariff protection were invested back into the industry, creating employment. The Democratic Party represented the big southern cotton, tobacco, and plantation growers that opposed tariffs because of their dependence on exporting goods. The battle between Republican tariff supporters and Democratic anti-tariff supporters dominated American politics from the buildup of the Civil War until the Great Depression.

Carver did not fully jump into the nation's greatest debate but was slowly pulled in. Carver had been admired and had his friend, James Wilson, in the great Republican administrations of William McKinley, Theodore Roosevelt, William Taft, and Warren Harding. While visited and courted by these administrations, Carver had stayed out of politics where Southern Democrats controlled the South with an iron hand. Neither Republicans nor their tariff policy were popular in the South. It was Carver's peanut research that now pulled him in. His development of milk from peanuts had caught the interest of peanut growers in 1919.

Carver's work with peanuts had resulted in an invitation to speak at the 1920 convention of the United Peanut Growers Association in Montgomery, Alabama. There were serious doubts about having a black speak to the convention.[2] The convention was at the Exchange Hotel where the Confederate Congress had met. Carver would enter through the back door. He had to take the freight elevator to reach the convention floor. Carver's presentation, "The Possibilities of the Peanut," and endless samples and exhibits won over all members of the association. He exhibited 145 peanut uses and applications, including milk and peanut flour as well as a wide variety of wood stains from the peanut. The group moved to make him a consultant and offered financial help in securing patents and the promotion of his writings. Articles by Carver soon appeared in *Good Health*, *Ladies' Home Journal*, *Liberty Bell*, and *Popular Mechanics*.

The United Peanut Growers Association was at the time in the midst of a major struggle to increase peanut production after World War I. The market was suffering from lower demand of peanut oil after the war and a surge of imported peanuts from China and Japan. In 1920, these Asian sources had captured over 50 percent of the American market. China, with its cheap labor for picking, was underselling American peanut growers by four cents a pound. This cheap foreign labor was compared to the higher labor cost paid (on an international level) by southern black farmers. America had over 300,000 workers employed in peanut production in 1920 (mostly black southerners) and 600,000 acres in production. Carver had led an explosion in peanut production that took the black sharecroppers from poverty. His uses and growing techniques for peanuts changed everything. In 1900, small farmers had received eight to ten dollars a month with a house to live in, a garden plot, and rations consisting of four pounds of meat and a peck of grain.[3] Peanut growing had elevated these farmers to three to four dollars a day, a house, a garden plot, and free fuel. They were on par with northern factory workers. The problem was that Chinese farmers were making sixteen cents a day! This difference would destroy the hard-fought economic gains of southern sharecroppers against racism and the boll weevil. Carver was as committed as the association in protecting American producers with scientific tariffs, the purpose of scientific tariffs being to equalize wage differences in other countries. Congress had a standing committee to calculate those differences known as the Tariff Commission. Higher wages were considered by the Republicans to be a result of American sacrifices to maintain freedom.

The political battle over tariffs in 1920 was one of the broader fights to extend Republican tariff protection to domestic industries. The Republicans had suffered political setbacks on their tariff policy under the Democratic administration of Woodrow Wilson, and many domestic products were in trouble. Southern Democrats had traditionally opposed tariffs because they were believed to have hurt southern plantations of cotton and tobacco, which were major exporters. Other southern crops were considered of little importance to King Cotton and tobacco. They also believed it increased the price of farm equipment. Northern Democrats opposed tariffs because they believed tariffs produced a general increase in food, clothing, and other goods. Republicans' protection of domestic industries had been consistent from Abraham Lincoln on. At the time, 30 million bushels of peanuts were imported from Asia, which represented about half of the American market. In 1921, the Fordney-McCumber Tariff Bill was in committee where rates were being decided on a long array of products. Peanuts would be of particular interest to the association, and Carver's dramatic and successful speech at their convention made him the ideal candidate to address Congress.

It would be no small feat to have a black man as an expert witness, but Democrats figured it would actually help their anti-tariff stance. No member of the committee knew of Carver at the time, and in general, a black man was unlikely to be a creditable expert with the Southern Democrats. While Southern Democrats mocked Carver, he would soon prove up to the task. He arrived in Washington two days earlier and was able to observe the lack of decorum he would face as a witness before the committee. It would be an adept and amazingly popular speech for this shy professor. Biographer John Perry offers a powerful picture of Carver's testimony:

> On January 20, 1921, Chairman Joseph W. Fordney, a Republican from Michigan, gaveled the Ways and Means committee to order.... The small black man settled himself into the seat at the witness table and asked if he could make room for some of his exhibits he had brought ... as previously bored committee members leaned forward in their chairs as Professor Carver began unpacking his box.... The chairman asked slyly if Carver had brought anything to drink — a reference to Prohibition, which had been in effect for little more than a year. When Carver answered that liquid examples might "come later if my ten minutes are extended," the committee broke out in laughter.... From sweet potato we get starches and carbohydrates, and from peanuts we get all the muscle building properties. Here one of the committee members interrupted the witness.... "Do you want a watermelon to go along with that?" A grinning colored man eating a watermelon — and showing lots of white teeth — was the stereotypical image of the time.

Carver didn't miss a beat. "Of course, if you want a dessert, that comes in very well, but you know we can get along pretty well without dessert."[4]

Carver continued to show the congressmen samples such as a peanut bar that used sweet potato syrup as a binder. His mock oysters made from peanuts proved popular with the congressmen.

Carver's peanut milk was also popular. He stated to a reporter, on this day seventeen years before he had met Henry Ford, "My very latest work in investigation is in reference to production of peanut milk. Like Mr. Ford I believe the cow is the crudest milk machine in the world and may easily be abolished for the synthetic cow, which gives clean, sweet, and wholesome peanut milk."[5] Carver talked of whipped cream, ice cream, butter, margarine, and coffee creamers from peanuts as well. Carver also demonstrated thirty different dyes. He demonstrated that peanut coffee was better than other popular grain coffees such as Possum.

As Carver proceeded to get extension after extension, it was clear that he was winning the day for the peanut growers. Democrat and future vice president under FDR, John Garner, tried to pull Carver into the legal points of a tariff, which he adeptly avoided. Garner argued that the tariff was but another tax and that the dairy lobby wanted to tax the margarine producers out of business. Carver would agree that a tariff was a tax, stating, "Yes sir, that's all the tariff means — to put the other fellow out of business,"[6] which drew loud laughter from the committee. After that comment, the chairman told Carver that his time was unlimited as he continued to pull samples out of his pocket and fend off Democrat attacks. He plugged his sweet potato throughout the presentation, all the while quoting liberally from the Bible. Carver, who was first allotted only ten minutes, spoke for an hour and forty minutes. In the end, an amendment was added to further increase tariffs on peanuts and vegetable oils by gaining the votes of some Democrats. The peanut growers got the tariff proposed of three cents a pound shelled and four cents a pound unshelled, which greatly increased peanut production. The Republican Party made major gains in the South with these new tariffs.

Carver's role in passing the 1922 tariff made a major difference in the long run. Democratic president Woodrow Wilson vetoed the bill, but Congress forced President Calvin Coolidge to sign it in 1922. Peanut acreage increased by 37 percent from 1928 to 1938 due in part to these tariffs, which Republican administrations increased in 1929. The uses of peanuts increased and peanut oil became an alternative to vegetable oils. The May 1921 issue of the trade magazine *Peanut World* summed up Carver's success:

> With profound pleasure and pride we dedicate an entire page to the incomparable genius to whose tireless energies and inquisitive mind of the South and country owe so much in the development of the peanut trades, arts, and industry. His contribution to the common fund of human knowledge in the field to which he has devoted his life is simply immeasurable. In the mental laboratory with which he has been so generously endowed he has been enabled, through the physical laboratory, scientifically applied, to be a benefactor not only to his own but to all races. He has been virtually a miracle worker.

Carver's performance caught the imagination of a nation and the hatred of many in opposition of tariffs. The tale of a former slave boy rising to lecture Congress was great copy for the press. The story was carried in papers across the nation. Carver himself would be given a syndicated newspaper column, "Professor Carver's Advice." Carver was a media gold mine for all, bringing visitors pouring into Tuskegee and financial backing. E. E. Sparks, president of Penn State, visited Carver at Tuskegee in 1924 to discuss his research with dyes made from soybeans and sweet potatoes. A representative of Mahatma Gandhi visited Tuskegee, and Carver recommended milk made from soybeans (grown widely in India) as help for the then ailing Gandhi. Interviews with Carver continued on an almost daily basis from other academics and the press. Fame brought enemies, too. Carver's constantly giving credit to God, acknowledging God's hand in all his discoveries, and quoting of the Bible made him a natural enemy of the evolutionists. At the time evolution was also a hot political debate, and while Carver rarely commented directly on the issue, his strong biblical stand implied that God was in control, not evolution. This unassuming man was collecting enemies because of his faith and color.

Carver became a popular speaker on the national scene, further aggravating racists, university scientists, evolutionists, and Southern Democrats looking to attack his creditability as a scientist. Militant black organizations opposed his concept of total self-reliance for poor black farmers. On November 20, 1924, Carver gave a long speech at the historic Marble Collegiate Church on New York's Fifth Avenue, the very belly of eastern liberalism and anti-tariff sentiment. He talked of God, Jesus, and the Bible being part of his laboratory because other than the Bible, "no books ever got into his laboratory." He talked of "inspirational" methods versus scientific ones. The talk resulted in the most famous attack on him by the *New York Times*. The headline was "Men of Science Never Talk That Way." The attack seemed to attack Christianity and Carver's credibility as a scientist, which played to enemy political forces.

Carver did draw large support in his letters to the editor. More importantly, they resulted in his most eloquent defense of the coexistence of Christianity and science: "I regret exceedingly that such a gross misunderstanding should arise as to what is meant by 'Divine inspiration.' Inspiration is never at variance with information.... I receive the leading scientific publications." He went on to give practical examples of how research was a type of inspiration. He would trace it to God, understanding that there are some "scientists for whom the world is merely the result of chemical forces.... I do not belong to that class." He further argued that he believed his information and inspiration came from a source greater than himself, the only argument being the ultimate source of information or inspiration, not magic versus science.[7]

Maybe even more to the point is that Carver didn't draw lines around religion and science. It was the view of most Victorians and today's evangelicals. For Carver, there was one body of knowledge. Henry A. Wallace, former vice president and secretary of agriculture, gave national support to Carver. Carver said of Wallace, "Few men combined the scientific and religious world so fruitfully as Carver. He never knew where one left off and the other one began."[8] Coming from the Bible Belt Midwest, Wallace understood it, but he also knew that few others would. The very deep and biblical faith of Carver was one of the differences with Henry Ford. Actually, it was Ford who had a more mystical view of the universe. Ford, however, showed a great respect for Carver's faith, even though he didn't fully understand it. Many years later Henry Ford would ask Carver to speak at Greenfield Village's chapel. Carver talked of God and about nature and communing with nature. This was where Ford and Carver found common ground, as they did on moral-based education. Ford's own faith grew only in his later years after he put aside the administrative duties of Ford Motor.

On December 30, 1918, Henry Ford (at age fifty-five) resigned as president of the company to focus on building up "other organizations," but he had one last car project to finish. It would be the building of the world's largest and fully integrated auto plant on the River Rouge. Henry was nearly done in his effort to take 100 percent control of the company, pushing out internal rivals such as the Dodge brothers and James Couzens. Now Henry could leave smaller matters to his son without worrying about internal interference while he was chasing other interests. To fend off another aggravation of the press, in 1919 he purchased a newspaper, the *Dearborn Independent.* His massive River Rouge factory would be built in Dearborn to avoid any interference from the city of Detroit. Finally, Henry felt he was in control enough to allow his son Edsel to take over day-to-day operations.

Henry A. Wallace, secretary of agriculture and later vice president for F. D. Roosevelt (Iowa State University Library, Special Collections, Box 9, Photograph 525).

The Rouge River as the location of a massive plant seemed a strange pick; however, Ford had his reasons. Besides avoiding city control, the location would allow Henry to use his electric motorboat from Fair Lane to manage the building of the factory. Still, the decision would require a costly dredging of the river to allow a deepwater port for Great Lakes ships. The Rouge River plant would be a megafactory in a farm field. The River Rouge plant would manufacture its own steel, iron, glass, and plastics. It would be the only fully integrated auto plant ever built. Ford would use his industrial architect, Albert Kahn, to design a complex of linked factories such as blast furnaces, steel plants, iron foundries, and glass factories. The Rouge would have the capacity to produce 4,000 cars per day from raw material to a finished car. The compound would have ninety-three buildings, employing over 75,000 employees. The plant required 300 guards alone to check employee identification cards. Its electrical generation plant could service all of southern Michigan, and the Rouge did sell excess electrical power. It had hundreds of miles of railroad track. Ford synchronized the transportation system of railroads and overhead conveyors to create a massive assembly operation.

The dream of the fully integrated factory on the Rouge River began in 1917, but

8. Massive Assembly, Peanuts, and Aircraft

Ford, as always, planned to do it his way. Ford's vision included glassmaking, steelmaking, an iron foundry, a national railroad, machine shops, a plastic molding department, and a massive assembly plant. Ford also planned to launch the building of the world's largest factory with almost no loans from banks. Some of the money came from Undersecretary of the Navy Franklin D. Roosevelt to build submarine chasers for World War I. The navy helped open up the Rouge River for deepwater ships to supply the Rouge. The Rouge operation would evolve over eight years of building, allowing Ford's managers to perfect new assembly methods. Ford visited daily, like Pharaoh overseeing its general construction, as his managers hammered out the details.

Ford overcame every problem with bodacious solutions. When faced with a shortage of cement, he railroaded it in from New Jersey. He used Thomas Edison's cement casting system to expedite building. Ford also needed to build houses quickly for his growing workforce, so once again he applied the Edison cast-cement method. The homes were considered ugly looking, but Ford was proud of the lumber-saving Edison approach. The Edison method used cast iron molds instead of wood framing for the cement. The cast-iron mold could be used over and over while lumber was a onetime use and waste of lumber. The war, however, prevented the building of these cast houses.

Ford wanted self-sufficiency in his manufacturing, and he demanded that his managers supply it. He had started paying thirty cents per square foot for plate glass; as his success pushed demand beyond supply, the price rose to $1.50 per square foot. In response, Ford added glassmaking to the Rouge operation. Furthermore, he automated the glassmaking process, driving the cost down to twenty cents a square foot. Ford did the same for steelmaking at the Rouge. In fact, he created more cost reduction, since auto making was one of the largest sources of recyclable steel scrap. Ford even took the idea of recycling old cars to a high at the Rouge. Dealers bought old cars, helping the buyer to purchase a new car. These old cars were sent to the Rouge, which had two production lines to deal with them. One line would overhaul the newer cars of the lot for resale. The others went to a disassembly line, which disassembled and sorted parts for various scrap. Ford demonstrated an amazing system of overall recycling and reprocessing that is once again getting looked at by manufacturers.

Model T final assembly production remained at Highland Park for years as the Rouge functioned as a supplier city for all the parts and subassembly. Ford did move tractor production to the Rouge, and in 1928, the Model A final assembly line was at the River Rouge plant. But even as Ford was building this massive supplier city at the River Rouge, he was considering a string of village manufacturing plants. Ford believed that the rural setting was better for productivity, and he carried proof of it. In 1920, Ford had opened a water-powered machining operation at Northville, Michigan. Northville would be the first of the Ford village factories of converted old gristmills. Although the waterpower proved insufficient and Ford had to convert to steam, the Northville plant pioneered air-conditioning, which Ford thought important for productivity. Ford was proud of the fact that it took 1.26 minutes to machine a valve at a cost of 4 cents. This compared to 3.5 minutes and 9.5 cents at his Highland Park plant.[9] Ford was already talking of a series of village factories in Southern Michigan in 1922.

Ford took his vertical integration of raw materials to the coal mines and gravel pits requiring him to purchase the Detroit, Toledo, and Ironton Railway to ship the coal from West Virginia and limestone from Ohio. He built two large iron-ore ships to bring iron ore from his purchased mines in Northern Michigan. At the Rouge, the entire car was produced with the exception of the tires. And a few years later, Henry would even develop rubber plantations in South America to move into tire production. Ford made vertical integration pay, giving him a huge advantage over his competitors. He took a bankrupt railroad and made it one of the nation's most profitable companies. Instead of buying locomotives at $65,000 apiece, Ford reconditioned old ones for $35,000 each. Ford opened his own grocery stores for his employees to assure lower grocery bills and ran them so efficiently he made a profit. The employee stores expanded into shoes, coal, meats, flour, vegetables, seed, and fertilizers. His Highland Park commissary did over $7 million in sales in 1926. Years later, Toyota would be more impressed with the efficiency of Ford's grocery store operations than his assembly lines.

Ford continued his war on waste, demanding the production and sale of by-products. His coke plant needed for the blast furnaces produced 20 million pounds of ammonium sulfate. He used the coke by-product of benzol to run trucks in the plant. Ford even supplied benzol to Detroit-area gas stations at a subsidized price for fuel in Model Ts. The coke plant also produced combustible coal gas for power generation. Ford distilled seven tons of garbage at his River Rouge plant daily to make alcohol, rivaling even the greenest manufacturers of today. He had a paper mill that used scrap wool, cloth, and paper to manufacture cardboard boxes. He used residue slag from iron and steelmaking to produce cinder cement blocks (many of which are still part of Detroit homes today). To keep his blast furnaces and steel plant running efficiently, he produced structural steel beams to sell. He recycled glass for his glassmaking operation. Instead of burning off residue blast furnace gas, Ford used it to produce steam and electricity to run his machines. At the time, part of the Model T frame was wood, and Henry Ford demanded that no waste be the rule of his sawmills. The loss of trees for bird habitat was a major concern for him. For cost-savings calculations, Ford demanded his accountants take it to one-thousandth of a cent. Using this approach, a simple fractional-cent savings in a single bolt for the Model T saved a half million dollars a year. Like Carver in agriculture, Ford made recycling and by-product use an objective of his industrial operations. The Rouge plant had an experimental lab to explore the uses of farm waste products.

Ford would always be torn between the farm and factory. His fortune would be in the factory, even if his heart was in the farm. Years later, it became apparent there was a pulling from both as he went into major farming projects. A reporter for *Fortune* magazine was not sure whether Ford was a "farmer with a bent for mechanics" or "a mechanic with a bent for farming."[10] Ford had a deep love for nature. Furthermore, he had a dream of bringing agriculture and industry together. He believed that there was a harmony that allowed him to build a huge factory in the rural setting of Dearborn, Michigan. His love of folk dancing, tractors, and plant research showed he probably was a mechanic with a bent for farm life.

Ford was a true conservationist also. He was driven to save lumber and American forests and to protect bird habitat. He argued, "The tradition of lumbering is of waste — that is why wages are low and prices of lumber high."[11] Ford applied strict conservation practices to his lumbering operation in Northern Michigan around Iron Mountain. Competing loggers had destroyed most of the Northern Michigan forest, but Ford improved practices to assure reforestation of his properties. He cut trees clean to the ground to favor regrowth versus others who left two-foot stumps. He mandated replanting and tried experiments in reforestation. He cut only mature trees, leaving younger ones for future years. It was at Iron Mountain that Ford started to develop his community-based approach to manufacture. Iron Mountain was a deserted mining area. Ford was proud of reviving the town, putting 5,000 to work at six dollars a day. The lumberjack camps had clean bunkhouses, laundering services, and well-prepared meals. Recreation halls were built for leisure. Running water, steam heat, and electricity were available at the larger camps. At the main town, Ford had a huge powerhouse built with steam turbines to generate electricity. In addition, he dammed the Menominee River with water turbines to generate electricity for the area.

Ford built a world-class wood distillation plant to produce alcohol. He used the latest technology so that any type of cellulose such a chips, bark, small branches, brush, corncobs, nutshells, and so forth could be used. In recent years, this cellulose alcohol production is again making the headlines as an alternative fuel. At his sawmills, like his factories, Ford invested in by-product production to reduce waste. Distillation of waste wood created an array of products. Every day Ford produced

> $11,000 worth of value from mill waste, including 125 pounds of acetate of lime, sixty-one gallons of methyl alcohol (one-fifth of the nation's total production), antifreeze, artificial leather, and fifteen gallons of tar, oil, and creosote. Sawdust, under brush, branches, wood chips, and cull lumber — that is, defective logs pulled from otherwise serviceable timber — were turned into charcoal "briquettes" (which today continue to be sold under the name Kingsford) or burned to power steam engines or heat worker bunkhouses.[12]

Ford got 600 pounds of charcoal out of every ton of waste wood as well as 600 cubic feet of fuel gas to run his steam engines. Many of these by-products were further processed to produce chemicals needed for auto production. Tar was processed into pitch and wood creosote.

Ford's conservation of forests remains a model today. Ford was proud of his forest efforts in Michigan and marked his land with Ford Motor signs. He built picnic areas and nature trails for travelers in the area. Interestingly, years earlier Carver had been arguing for the conservation of wood in the South in a 1902 article, "The Need for Scientific Agriculture in the South."[13] Conservation would be one of many things Ford and his future friend Carver would see eye to eye on. Years later standing with Henry Ford in Ways, George Carver said, "Conservation is one of our big problems in this section. You can't tear up everything just to get a dollar out of it without suffering a result. It is a travesty to burn our woods and thereby burn up the fertilizer nature has provided for us. We must enrich our soil every year instead of merely depleting it. It is fundamental that nature will drive away those who commit sins against it."[14] Ford and Carver saw wood and forests as precious resources for industry and agriculture as well. In his bulletin number 6 (1904), *Feeding Acorns*, Carver argued that acorn feed could help save mighty oak trees from the axe of timber companies.

Ford's conservation went far beyond his own manufacturing operations. In Michigan, Ford purchased many acres of "cutover" where greedy lumberers had harvested the forest leaving stumps, making it unusable for wheat farming. Ford lamented the destruction of Michigan pine forests. In many ways, he would become the "Carver of the North," reclaiming land for farms. He bought thousands of acres in Clare County, Michigan, near Harrison, to prepare the land for farming. Ford used an army of tractors in the 1917–1919 period to take out the stumps. The stumps were used as fuel or in alcohol production. Sometimes it took years to clear this land. In the intermediate clearing phases, Ford planted potatoes to help improve the fields. Eventually, Ford created fully operating farms where the land had been declared unusable.

At the Rouge plant, Ford had a "wood salvage department" that recycled and reprocessed all scrap wood. Ford had an all-company rule that wood crates and boxes were to be opened with care, so the wood could be reused. Workers removed nails and spikes, sanded, and cut the wood to usable sizes. Ford built a box plant to run off of the recycled wood. He estimated he saved over 100 million feet of wood a year. Special presses and dies were built to process smaller pieces of wood to the larger sizes needed. Ford manned the department with "substandard" men who, because of injury, could no longer man the assembly line or do heavy labor. Ford argued he saved millions of dollars, but some of his managers argued these salvage operations costs were based on simple cost accounting of the time. Ford, however, always looked to the bigger picture in his analysis of savings. He restrained his managers from building wasteful practices of any resource into his manufacturing operations. After Ford's death, these wasteful practices crept back into the operation, only to be taken out again in the 1990s by Japanese industrialists based on the original work of Henry Ford.

Ford not only wanted to reduce waste but pioneered new materials from the farm in his auto production. He was every bit the industrial equivalent of George Washington Carver. Ford spent years perfecting the production of artificial leather. By 1926, he was processing cloth with nitrated cotton, ethyl acetate produced at his wood distillation plants, castor oil, and benzol. This artificial cloth was in many ways similar to later rayon products. The product was then pressed and heated. Ford replaced wood in steering wheels with a product called "Fordite" made of straw, silica, sulfur, rubber, and soybean resin extruded in rods. The rods were formed into the steering wheel and baked to a hard part. Ford's work with leather and upholstery in the early 1920s led to his experimental work with plants. In particular, he put his scientists on using flax-based cloth to replace cotton. He produced artificial wool from soybeans. He reasoned that flax, popular for the manufacture of linen in Europe, could be grown in Michigan to reduce transportation costs. He then set up linen mills to produce it. The experimental work with flax went on for years in an effort to reduce the high labor costs of flax production.

Like George Washington Carver, Ford was thrifty in all phases of his life, albeit at a higher society level. Ford increased his travel and outside projects during the 1920s. To facilitate this, Ford built a private Pullman railcar. At the cost of $159,000 in 1921, it was considered the best-furnished private railcar in the world known as *Fair Lane*. Yet Ford was often renting railroad cars at $5,000 a week. The roof was built out of an expensive corrosion-resistant alloy of nickel-copper know as Monel (close to the composition of today's nickel coin). Several presidents would travel with Ford in the *Fair Lane*. It could carry, feed, and sleep twelve persons on cross-country trips. He used it as a traveling hotel. It was walnut paneled with a full kitchen to support large dinners. Henry made annual summer railcar trips to his timberlands in Northern Minnesota. Clara Ford used it for shopping trips to New York, while Henry used it for trips to his Ways, Georgia plantation and his summer home in Fort Myers, Florida. For summer vacations on the Great Lakes, Henry added luxury quarters to his ore ships. Each summer Henry and Clara made a five-day lake excursion to Northern Michigan to pick up iron ore for his Rouge factory.

Ford looked at the workers of his factories as a type of raw material, and there were many savings to realize with better management of the human element. Ford expanded again into social engineering at the Rouge operation and developed new community manufacturing arrangements. His sociological department at the old Highland Park plant was never popular, and many managers such as Charles Sorenson had resisted its full implementation at the River Rouge plant. Ford turned to much broader utopian views. He wanted to build an industrial/agricultural complex for Southern Michigan. The River Rouge plant itself was set up as decentralized manufacturing cells feeding a grand assembly line. Ford wanted to build a network of farm village factories to feed his assembly plants in the 1920s.

With his Northville valve plant, Ford looked to decentralize more into Southern Michigan so as not to create an urban city in Dearborn. Ford was careful to assure that the Dearborn area had sections of wilderness in an effort to minimize the impact of his huge industrial core which used nearly fifty train carloads of coal every day. Henry Ford truly believed that industry and the environment could be integrated to function in harmony. Ford would live his whole life on his Dearborn game preserve, Fair Lane, which today remains a beautiful park on the Dearborn campus of the University of Michigan. To Ford's credit, the environmental balance he worked on in Dearborn can still be seen today. This concept of a utopian working community would lead to other projects such as Muscle Shoals, Alabama; the Village Industries of Southern Michigan; his Ways, Georgia, development; and his idealized rubber plantation — Fordlandia.

The 1920s were a rush of creativity and projects for Henry Ford. In business, Ford purchased the assets of Lincoln Motor Company in 1921. Lincoln Motor had roots going back to the old Henry Ford Company of 1902. His old partner and friend, Henry Leland, owned the company at the time of bankruptcy. The Lincoln was a top-of-the-line car, and it was considered part of Detroit's industrial legacy. Ford was considered a community hero when he purchased the assets and resumed car manufacturing at the plant. Henry would turn the operation over to his son, Edsel. Henry now had the money to dream at new levels. Ford would launch what seemed to be an endless array of industrial, social, and agricultural projects. Early on, Ford expanded his Dearborn farm production. In 1917, he built grain storage silos and a grain mill. The grain mill used novel grinding rolls to produce very fine flour. He sold the flour to his employees and started bakeries to make bread for his employee stores. Ford even developed recipes for bread using some potato flour mixed in. The use of potato was favored because Ford had invested in Michigan "cutover" land, where potatoes were planted.

Ford proved himself a pioneer of new methods and an endless experimenter in both industry and agriculture in the 1920s. He had already taken vertical integration in manufacturing further than anyone prior. Vertical integration had given Ford huge advantages in costs and consistent supply. Still, the speed at which the River Rouge and Highland Park plants could make cars continually tested the operation of the supply chain. Shipping delays, rail bottlenecks, and slow rail connections became the weak link in his massive supply chain. Ford moved aggressively into rail and shipping transportation, promising to bring his efficient and lean management techniques to these industries.

He seemed inspired that reporters were calling his successful operation methods "Fordism." Even in Communist Russia, Ford was hailed as the world's greatest efficiency expert. Ford purchased the bankrupt Detroit, Toledo & Ironton Railroad in 1922 almost to prove a point. It gave him seventy steam locomotives (only fifty were working) and 2,800 railcars. It gave him 456 miles of track through Ohio and the supply chain to the Ohio River, where it could also pick up more iron ore and coal from West Virginia. Ford ran the tracks directly into the River Rouge, making it the world's greatest supply chain of raw materials in the world. Ford became known for turning around a bankrupt railroad and making it the country's most profitable. He was only restricted by the federal government, which required that half the profits beyond 6 percent be redistributed to all the other railroads in the country.

Part of that turnaround was improved engine maintenance. Ford also repaired and added bridges to allow for maximum speed and also made reduction of waste the railroad's mission. Ford said in 1926, "It is said that a D.T. & I. man always carries in his hand a bunch of cotton waste [for cleaning]. It is the insignia of the road. But cotton waste is not thrown away once it is used. It goes back to a cleaning plant and comes out as new. No scrap is thrown away. It all goes back to our reclamation plants."[15] Like his other projects, Ford could not resist the urge to experiment. He built a gasoline railcar to use on his railroad. The single gas-powered railcar called the *Dearborn* was not for personal use, although Ford and his son Edsel did drive it.[16] The idea was to develop a gasoline streetcar for use in the city of Detroit. For whatever reason, the railcar died an unheralded death. In addition, Ford added an electric railroad link from Flat Rock, Michigan, to Dearborn. His successful and innovative foray into railroading came to an end in 1928 when the government brought suit against him under a 1903 railroad act. The government charged him with a lack of proper rate charges to his internal operations! Instead of fighting and dealing with government regulators, Ford sold the country's most profitable, highest-wage-paying, most successful railroad in the country.

More significant during the 1920s was the birth of an idea for a historical village. Ford's development of an industrial and agricultural museum would prove just as audacious as his assembly line. Henry had been collecting farm equipment for years, but now he moved into restoration projects. While other industrial barons collected art, Henry looked to collect American culture. In 1923, he purchased the old Botsford Inn on Detroit's

Grand River Avenue for $262,760. He and Clara had danced there in their youth. The Botsford Inn was built on an ancient Indian trail and had been a stagecoach stop and toll gate since 1836. Henry restored it and added vintage farm buildings to the property for $399,483. He completely restored the large dance floor and searched the country for fiddlers. For dancing, he added nine rare violins for $103,200. He hired a dance instructor, Benjamin Lovett, and set up a dancing school at his Dearborn Engineering Building. The same year, Henry rescued the famous 1688 Wayside Inn in Sudbury, Massachusetts, which had been immortalized by Henry Wadsworth Longfellow in "Tales of the Wayside Inn." In restoring these inns, Henry added collecting of early American artifacts to his farm equipment hobby. Actually, the first restoration had been Henry's birthplace and farm in Dearborn.

By 1924, Ford's artifacts were filling up the old Fordson tractor factory in Dearborn. Henry had restored a huge Westinghouse steam engine he had operated back in 1881. He was similarly following his youthful activities by collecting clocks and watches. Ford had also launched a nationwide search for rare editions of the *McGuffey Readers*. Clara was stockpiling old rugs, china, pewter, and other Americana. The collections grew to where the Ford Airport hangar was storing the overflow. Ford announced in 1925 his plan to build an educational and historical village about two miles from Fair Lane. He talked of a "village" that would show and tell the story of manufacturing and crafts. Combining two dreams, it would also have a trade school for young boys. The location consisted of 240 acres across from the Dearborn Ford Engineering Building. His historical and educational village would have to wait, however, as Ford would follow Plato, Thomas More, H. G. Wells, and Jules Verne in the hope of building a utopian society.

9

A Better Life — Industrial and Agricultural Utopias

> "The American idea of business is based on economic science and social morality — that is, it recognizes that all economic activity is under the check of natural law."
>
> — Henry Ford, 1926

In the later 1920s and early 1930s, Henry Ford was free to pursue his dream of an educational and historical village, but first he would try to establish an industrial utopia. Carver would continue to search for ways to improve the lot of poor southern farmers. Both men would move beyond technology to the more difficult social and cultural blocks to their dreams. Both Ford and Carver changed the culture and economy of the nation, but they came from two very different existences. Ford traveled in the company of princes, kings, and presidents on his private Pullman railcar. For Carver, fame had brought a surreal world of traveling in color-segregated railroad cars and entering the back door to speak at major events. Carver persisted in his mission to improve the life of black farmers, including the improvement of race relations. He realized that his economic improvements would mean little for poor black farmers without improved racial relations. Ford had welcomed the black man to his factories. Both men would change their country for the better by educating it, because for both, America's future was in its ability to allow its best creative minds to create. Carver had to educate a nation on the ability of his race to create, while Ford had to break the structure of education that inhibited creativity. Both men faced many roadblocks in their midlife missions. Finally, both Ford and Carver saw agriculture and industry as linked and symbiotic. A great nation, in their minds, needed both. They both sought to bring the founding division of Jefferson and Hamilton together in a new nation of interrelated agriculture and industry.

After his talk before Congress, Carver was in great demand for talks and lectures, which changed his life and gave him a role in the improvement of race relations. One biographer said, "By the mid-1920s Carver was a celebrity on a par with movie stars. Some of the lectures were invitation-only events attended by society types."[1] Still, recognition with whites in the South would come slow. Furthermore, Carver's dreams of commercial success eluded him, but he found success in improving the economic lives of his region. Carver's power as a speaker and product marketer brought new interest in his own work and inventions. Carver's own hopes had been dashed as his paint company had failed and others had patented his better ideas such as peanut milk. Similarly, Carver had hoped that his diabetic flour from peanuts and sweet potatoes would find a market, but it failed to do so. Many writers had argued that Carver had little interest in money, but that appears only partially true. Financial success was a measure of scientific success in Carver's time. It was Edison's commercial success that gave him creditability even with his lack of scientific procedures. In fact, Carver's lack of commercial success is still used as a criticism of his scientific contributions. Carver's personal ventures failed, but Tuskegee received ample support for Carver's laboratory. One of Tuskegee's local contributors, Ernest Thompson, had always followed Carver's experiments with great interest. Carver's thrust onto the national scene in 1920s offered Thompson new hopes of raising capital for a business based on Carver products such as paints, dyes, and stains. Thompson proved a poor investment manager and Carver a poor promoter. Some failure was due from major companies questioning the creditability of Carver as an industrial chemist. Carver often used his own lab terminology which confounded university-trained chemists. Carver preferred to use descriptive process formulas versus chemical notation. For example, he called his outstanding Prussian blue dye a sex-triple oxide because he developed it through six oxidizing steps.[2] The commercial development was further limited by Carver's secretive approach in not revealing his formulas. Carver had already lost a few patent battles and knew it would be hard to win against corporate lawyers and chemists. Still, many of Carver's recipes, inventions, and applications would play a role in the economic development of Alabama.

Thompson and Carver suffered a big setback in their hopes to sell products to Sears Company, which served on the Tuskegee Board of Trustees. Sears felt Carver lacked the factory capacity to supply routinely. Commercialization of Carver's products did gain some momentum again in 1923 when Thompson put together a number of investors. He had seen many of his ideas go commercial from his various talks, and Ernst Thompson looked to use Carver as his salesman at exhibits and fairs throughout the area. With the help of the Atlanta & West Point Railway, Thompson set up a Carver exhibit of products at the Cecil Hotel in Atlanta. The railway sent a special car to pick up Carver and his exhibit. It was a beautiful rooftop garden exhibit. It included the many uses of the peanut and sweet potato. Another part of the exhibit included paintings by Carver using Alabama clay colors. The publicity of the exhibit was extremely positive, and progress with state investors moved forward. Carver moved toward the completion of patents on paints, stains, and cosmetics from clay and plants. The Carver Products Company was incorporated, and efforts were focused on the paints initially. Experimental trials of paint on railroad cars were promising, but capital never was fully developed to build the factories.

Carver's main hope for commercialization was in his paints. In 1924, the investment group purchased some clay deposits and land near Chehaw (about three miles north of Tuskegee) for $27,000. Carver noted his hopes in a 1924 letter: "I have always felt that the work on clay was really the biggest thing I have been able to do, as it is far reaching in its possibilities."[3] Carver had used the clay in face powder, ceramic paints, and wood stains. He had also used the clay to produce a high-quality glossy paper and plaster. Carver believed in his paints because he had successfully reduced painting costs at Tuskegee and nearby schools, farms, and churches. Carver had been making paint since 1911 and

Carver's traveling exhibit of Alabama clays for industrial applications (Bentley Historical Library, University of Michigan, File HS8582).

had given the recipes in his bulletins. The reasons for lack of manufacturing success appear to be related to finance. Carver continued to search for commercial applications of his products through the 1920s.

Being black and wearing his old suit presented some difficulties for Carver in winning over southern businessmen. His cornstalk bow tie didn't help in a world of businessmen. Carver's personal lack of business skills was part of the problem as well, but there were serious political problems with black economic progress in the South at the time. The white "Bourbon" Democrats had taken control of the South from the old Lincoln Republicans. The Bourbon Democrats claimed a new focus on southern industrialization, but they were also antiblack in their social outlook. In general, even this "New South" economic policy of the Bourdon Democrats was not very open to black enterprise. Carver's push for the small black farmers and southern industrialization was not popular with the Bourbon Democrats. Carver's biographer, Linda McMurray, observed, "Carver's ultimate goals were incompatible with those of the Bourbon Democrats."[4] Carver never talked politics, but he was in an era of highly charged black politics in the South. Furthermore, while Carver was apolitical, he had been associated with the Republican administrations

9. A Better Life—Industrial and Agricultural Utopias

of McKinley, Teddy Roosevelt, and Taft. In reality, it is difficult to discern the real nature of these commercial failures, but commercial enterprise was not one of Carver's strengths. In fact, Carver's interest in commercialization appears more a matter of creditability and acceptance from the scientific community than gaining wealth.

Still, Carver's exposition of products in Atlanta was a huge success. Carver was at his best in this type of show-and-tell venue. Clothes or not, Carver could speak and make great presentations. Some financing finally came to build a small paint factory, but the project never blossomed into a business. The company also tried to sell some cosmetic products based on clay and peanuts, which could be manufactured through similar processes needed for the paint operation. As a former trainer for Iowa State's football team, Carver believed strongly in the restorative properties of peanut oil for massages. Carver had even been recognized by the Tuskegee football team, as he was a great fan. Some still believe in the use of peanut oil because of its vitamin E content as useful in the treatment of arthritis. However, the peanut oil did not have the strong marketing programs of competing products. A decade later Carver did find some success with cosmetic products. "Carvoline," an antiseptic hairdressing, did find a small market niche. And in the 1950s, Carver's assistant, Austin Curtis, built a successful cosmetic line based on Carver's formulas.

Carver had some limited success with herbal and medicinal products in his lifetime. A mixture of creosote and peanuts was used in a medicine for respiratory problems known as Penol. The famous New York speech of Carver mentioned Penol, which gave it a publicity boost. Penol remained on the market for years as a small success. Carver continued to try to market cosmetic products and medicinal products with limited success over the years. Some investors tried in the late 1930s to market Carver's Peano-oil as a hair preparation and a "Miracle Massage Oil." Neither of these products seem to have made much money. Later in his life, he noted that this effort was based in vanity and took him away from his real mission in bringing in a "New South" with manufacturing and better interracial relations. Carver's impact in this mission was an unquestionable success.

His success and mission ran counter to the decline of racial relations in the South as Carver came to Tuskegee. The 1896 Supreme Court decision *Plessy vs. Ferguson* voiced its opinion of separate but equal. Using this, Bourbon Democrats put in a structure of segregation and a policy of social and economic discrimination. Carver's successes in Congress and fame offered a hope against the trend in the South. In 1923, Carver won a groundbreaking award that shocked many. The Atlanta Chapter of the Daughters of the Confederacy gave him a written letter of appreciation. However, a problem was a split in black political factions in the South. There were two rival groups. One was the "accommodationists" of Booker T. Washington and Tuskegee supporters, and the more aggressive, newly formed National Association for the Advancement of Colored People (NAACP). Some of the harshest critics of Tuskegee and Carver were in the NAACP. But even here Carver offered a visual bridge, when in 1923 he won the Spingarn Medal of the NAACP for his work in agricultural chemistry. The Spingarn was particularly satisfying and consistent with Carver's mission. It was given to highlight the achievements of black scientists to young blacks. The apolitical and successful Carver was the perfect catalyst for a real change in the nation. He faced racism with calm determination to succeed.

Paint-testing panels at Tuskegee (Bentley Historical Library, University of Michigan, File HS8576).

Carver realized that improved racial relations were a key component of his improved economic vision for the New South. Carver would be asked by the YMCA to participate in a lecture tour of white colleges to help build a new future for racial relationships. The tours would be funded by the Atlanta-based Commission on Interracial Cooperation, which was a moderate black group that had long been supporters of Tuskegee. The college tours started slow with visits to Mississippi State, Mississippi College, and Clemson, but Carver's ability to win over the crowd soon brought a flood of invitations. Other visits included North Carolina State College, University of North Carolina, Roanoke, Duke, and Furman, with Carver traveling over 2,000 miles by automobile. In addition to the YMCA, the Atlanta-based Commission on Interracial Cooperation (CIC) was an early movement on improved racial relationships. It had close ties to Tuskegee and had known of Carver's skills. Through the CIC and YMCA, Carver made many successful trips to white colleges to speak. The CIC and YMCA approach of Booker T. Washington was never popular with black activists, and Carver often was attacked from both ends of the spectrum. Still, Carver made steady progress.

Annually, Carver spoke at the Blue Ridge YMCA Conference, which spread the interest and the fame of Carver among the southern universities. In the early years of these visits and conferences, Carver was forced to sleep separately from the white attendees, but the barrier was soon eliminated. Still, during most of his trips, he was forced to sit in segregated railroad cars. Carver endured this racism while not losing sight of his mission. The humiliation of this great scientist was a statement in itself that did more to change minds than any protest. In the end, Carver's greatest contribution may have been in his role in improving interracial relationships in the nation. Over the years Carver formed a long-term letter-writing network of "Blue Ridge Boys." This network would leave a legacy of written letters in various university archives, which offer the best insight into the man.

In the 1920s, Carver made trips via railroad stops throughout the South. In a twelve-day trip through Virginia, Carver outlined his itinerary: "My first stop was made at Petersburg, made two in Richmond, one at Stanton, one at Lexington, at Washington and Lee, two at Lynchburg, and an all day talk at Emory and Henry. Emory Va. that is I first met the Sunday school, then little tots, went to church, dined, went out riding for an hour, then a volunteer round table talk in which I attempted to answer questions of many, many kinds for two hours and to a large audience, suppered, then a main lecture and of course talked afterwards to many who were interested. Some followed me to my room and talked until 11 P.M. so you see I had a very, very busy day."[5] Not surprisingly, Carver suffered from exhaustion and had to postpone a similar trip through Tennessee a few weeks later.

Carver's trips increased through the 1920s, and he often made two-week rail trips through the country to speak at Lions Clubs, Kiwanis, chambers of commerce, universities, fairs, expositions, and churches. These nonuniversity talks came with their own stress for Carver facing white southern audiences. Carver's farming and household tips quickly won over both white and black audiences. In addition, the Commission of Interracial Cooperation and the YMCA felt he could help lower racial tensions and used him as a speaker frequently. He became as popular with white groups as black groups. Unfortunately, he often was forced to travel in special colored cars, use separate restrooms, and was unable to eat at many restaurants. Carver took these put-downs stoically and persisted to inspire both whites and blacks. Black leaders often misinterpreted his constant prodding of young blacks to get an education and work hard. Much like Bill Cosby today, Carver went to the black problems before addressing the problem of racism. Many criticized his approach, but no one could criticize his effort.

He was well aware of racism firsthand but thought the individual had little control over it except through working to overcome it with hard work with stoicism. Racism surrounded the Tuskegee Institute as well. The Ku Klux Klan was still active in the Tuskegee area, and often a simple trip to nearby towns brought verbal attacks. Traveling, Carver had faced name-calling and threats at every white college he visited. However, he faced the racism with stoic determination. It had been Carver's own personal reaction to racism since his youth. There is no doubt that Carver's greatest contribution to ending racism was his very success in science. In 1923, even the biased *Atlanta Journal* had credited his "amazing scientific genius and soul of a dreamer." The same year, *Success* magazine called him "Columbus of the Soil."

Carver continued his research work through the 1920s. He found similar interests to those of Henry Ford in Dearborn. In 1925, Carver asked the government for research funding for the production of alcohol from plants. He also requested possible funding to research flax as a substitute for cotton, another idea that Ford had also been interested in. Like Ford and Edison, Carver had even started to experiment with latex substitutes such as sweet potatoes as a source for rubber. He tried new plants and crops for the South, experimenting with making dyes from avocado seeds. He continued to exhibit his work for the promotion of the Tuskegee Institute as well. He put together a major exhibit at the Southern Exposition in New York in 1925. Alabama supplied some funding, and it paid off as Carver's exhibit drew large crowds. The exhibit highlighted many of Alabama's resources such as its clays and its potential for expanded agriculture. Alabama won the award for the best exhibit. More of Carver's exhibits at state expositions followed this success. It also created a national demand for out-of-print bulletins by Carver. Unfortunately, Carver's popularity and recognition as a scientist never translated into personal commercial success.

The 1920s brought new challenges to cotton growers as synthetic fibers such as rayon drove the price of cotton to new lows. In addition, the nation experienced a recession in the early 1920s that hurt industry and farming. What cotton the boll weevil had left, new materials and the Depression would destroy. Cotton stockpiles increased with large surpluses, so once again Carver looked to alternative uses of cotton. He suggested the use of cotton fiber as a binder in asphalt roads. Over the next decade, the use of cotton increased in road building. Carver also experimented with novel production methods of mixing cotton and asphalt at the ginning operation and producing blocks of material for paving. The paving would require forty bales of cotton per mile.

Carver also suggested a type of pressed wallboard from short-staple cotton, which was the main southern crop. This work led to other types of pressed and insulating boards. He produced a wallboard made of corn and cotton stalks and peanut shells. Other chemurgists such as J. Harris Hardy applied Carver's work to make cotton fiber roof shingles. Carver expanded his work on developing long-staple cotton for tire making. The automobile had driven the price of long-staple cotton to three dollars a pound versus twenty cents for short staple.

Carver experimented with wood chips, sawdust, and sweet potato glue to produce panels which approached the popular pressed boards of today. Many of these products would become the core of chemurgical research in the next decade. Carver was always on the lookout for ways to attach new industries to farm products as well as to improve the efficiencies of the farm. He continued to mix research into home economics in his writings. In hard economic times, survival depended on subsistence and home economics of the farmer to last the length of the recession. Carver's talent to write and communicate was why the government asked for his help during wars and economic downturns.

Carver's ability as a science writer, home economist, and do-it-your-self writer made his writings popular in the trade journals like *Peanut World* and *The Peanut Journal*. Carver exhibited the same popular style and added biblical quotes in many of the articles. He was allowed editorial freedom to add other studies such as those on pecans and sweet potatoes. He promoted the peanut as the "king of legumes," and the pecan as the "king of nuts." In addition, he used these magazines to promote the drying of fruits for southern farmers. He added recipes to further expand the uses of southern crops. He often commented on how a poor farmer could live completely off the land in times of great economic downturns. He suggested the use of newspaper to insulate walls better. His nationally syndicated newspaper article grew in popularity during the 1920s farm recession. He was now free to write without financial limitations on the publishing end that had restricted his Tuskegee bulletins. Carver was becoming known to men like Henry Ford, Harvey Firestone, and others through his writings.

Carver became an important prophet for change in the South. He wanted more than just subsistence for the farmer. Carver had experienced firsthand the wedding of the southern farmer to cotton. No matter how poor the crop or the size of insect plagues, farmers clung to cotton. Even when faced with starvation, black farmers chose to migrate north to the Ford factories rather than change crops. Carver remained relentless in his pursuit of new crops and industries for the South. Carver argued in 1922, "My general feeling is, that as soon as the South can readjust itself to the above condition, that agriculture will be greatly improved on account of more intelligent farmers coming in. Scientific investigation and demonstration will bring forth other money crops. Factories will come in. The South to my mind will soon become alive as to its vast mineral resources in clays, sand, etc. In time, the whole South and whole country will be helped."[6] Carver argued that industry and agriculture were interrelated. It was a vision that he shared with Henry Ford.

Ford had his own variation of the vision of the New South and the industrial-agriculture utopian community. Ford now had the money to do what he wanted as the world's wealthiest person. His personal wealth was estimated as bringing in over a quarter of a million dollars a day. Ford had started his quest for an industrial utopian community with his new factory design at Highland Park and the River Rouge plant. He had located the River Rouge factory in Dearborn to build not just a factory, but also a community and link to the surrounding farming community. His huge River Rouge plant was initially a diversified mega-grouping of small factories, initially supplying the Model T line at Highland Park, then a huge assembly line within the River Rouge plant. He had started a lifelong project of village industries on the Rouge River and other southern Michigan rivers to supply Highland Park and the River Rouge factory. Ford would predict in a number of media outlets in 1924 "the passing of big cities and the decentralizing of industry."[7]

But Ford's real motivation for improved communities through industrialization and agriculture came from the building of his Cork County, Ireland, factory in 1919. In 1912, Henry, Clara, and Edsel visited Ireland in search of the family roots in Cork County. Ford was appalled at the poverty of the county and donated $50,000 to the local hospital and $5,000 to the local Sisters of the Assumption to help the poor. It was the same poverty that had originally driven his family from Ireland. Ford decided to build a factory there in 1917 noting, "We choose Ireland for a plant because we wanted to start Ireland along the road to industry. There was, it is true, some personal sentiment in it."[8] Though started in 1917, because of the war and some local politics, it was not completed until 1919.

Prior to the Ford plant at Cork, a townsman could at best make sixty shillings (fifteen dollars at the time) a week from three days at the docks. Unemployment was probably near 40 percent. Most men had but one set of clothes. The Ford plant employed 1,800 men for five days a week for a pay of 100 shillings (25 dollars). This corresponded to his five-dollar-a-day minimum wage in the United States. The men now could feed their families, have two to three sets of clothes, and even have money for family entertainment. Ford, of course, imposed some behavioral requirements such as sobriety and not smoking on the job. For sure, these were difficult requirements for many. Ford also resisted the unionization of the men which was common throughout Europe at the time. He continued to give to community development as well. Ford was proud of how the higher wages had turned the very culture of the community around. Ford also learned from Ireland the use of flax linen as an improved alternative industrial fiber to cotton. His use of local agricultural products further helped the overall economy of Ireland. Inspired by his Cork and River Rouge plants, Ford dreamed of ever-bigger projects of decentralized industry and agriculture. He saw the great opportunity in the South that he had seen in Ireland.

The center of Henry Ford's interest was a thirty-mile stretch of rapids on the Tennessee River known as Muscle Shoals (nearest town) in northwest Alabama. For perspective, Muscle Shoals is about sixty miles from Huntsville, Alabama, 120 miles from Memphis and Nashville, Tennessee, and 200 miles from Tuskegee. Today it is part of the Tennessee Valley Authority of great dams. This part of the Tennessee River Valley was part of the southern Appalachian Mountains; and in 1922, it was home to over 4 million people, most of whom were in poverty. This part of the river with its swift rapids and falls had been an early block to navigation of the Tennessee River in the 1800s. The surrounding area was made up of small and poor cotton fields. As early as 1820, many sought federal aid to develop canals on this portion of the river for improved navigation for the cotton growers. Furthermore, annual flooding washed away topsoil, leaving yellow clay. By 1923, 7 million acres were considered too severely eroded to farm properly.

The building of the first American hydroelectric generating plant at Niagara Falls in 1896 had shown the potential of electrical energy to drive industry. President William McKinley had signed a bill in 1899 to start dam construction expanded throughout the country, only to have it overturned by President Teddy Roosevelt. The success of Niagara had created the interest of Tennessee and Alabama politicians to gain federal aid to utilize the power of the rapids and falls to generate electricity on the Tennessee River. The thirty-seven-mile stretch of river around Muscle Shoals dropped 134 feet in elevation. In 1916 with World War I looming, interest grew in the potential of Muscle Shoals to generate electricity and to produce nitrate fertilizer for cotton growers and nitrate for explosives and munitions. National defense supporters and "King Cotton" congressmen were able to pass the National Defense Act of 1916, which provided for the funding of hydroelectric generation and nitrate production at Muscle Shoals.

At the end of World War I, the government lost interest in the completion of the project, and Congress denied further funding. The government had already spent more than $105 million in the stalled project and was paying over $300,000 a year just to maintain idle installations. The assets lay wasting away with no hope of funding in 1922. Ford looked at the area as a potential industry-agriculture utopia. Besides paying for future construction, Ford would pay the government $5 million in cash and $1.5 million a year in rental fees. Ford's estimated fortune at the time was probably in excess of one billion dollars, so he clearly had the money for such a grandiose plan. In 1921, Henry Ford and Thomas Edison explored the Muscle Shoals area in search of its potential to be the "Niagara of the South." Thomas Edison said of Muscle Shoals, "The possibilities here are greater than all the gold in the world."[9] Ford already had experience in using hydroelectric power at his factories, and he looked forward to a world-class jump into power production. In addition, the country was in the great farm recession with cotton prices dropping from forty cents a pound to fourteen cents. Money was extremely tight, with most farmers unable to get credit from the banks. Farmers were burning corn and cotton instead of paying twelve dollars a ton for coal.

Ford envisioned a seventy-five-mile long city he would call "the Detroit of the South"—employing over a million people—in proposing that the government allow him to purchase this unfinished project. More importantly, Ford would offer his employees the five-dollar-a-day minimum wage of his northern factories. Most of them would be poor black farmers. Less than 2 percent of these farmers had electricity. A petition of the poor farmers in Ford's support called him "Moses for 80% of us."[10] Ford hoped to turn the area into an American utopia of economic and personal freedom. Ironically, in 1937 after Congress turned down Ford's proposal in 1923, George Washington Carver visited Muscle Shoals to speak at the local chapter of the American Chemical Society about Ford's initial vision. Carver was refused a room in the very hotel Edison and Ford had stayed at and at all other hotels in the area.[11] The Muscle Shoals area had become one of the poorest and most segregated in the South.

Ford's vision included a completed dam, two fertilizer plants, aluminum manufacturing, and steel manufacturing. Furthermore the nearby agriculture would be improved with not only nitrates but also a planned phosphate quarry. A series of two dams would create 1 million horsepower, the most powerful generating station in the world. It was to be the world's greatest engineering project beyond that of the Panama Canal. In fact, the scope of the overall project would rival great nations in history. Ford's plan was a masterpiece of urban, electrical, mining, agricultural, and social engineering. It was to be a city of diversified village industries supported by surrounding agriculture. Employment would reach 100 percent to the area with Ford's policy of employing anyone who wanted to work.

The core of the Ford proposal was two dams, one at Muscle Shoals and the other at Florence on the Tennessee River. The dams would form a seventy-mile narrow lake with a city and village industries along the shore. Ford estimated that the cost of building the dams was $45 million. One of the dams would be the completion of the government-started "Wilson Dam." The hydroelectric power would be used to operate two huge nitrate fertilizer plants, which again the government had started. Ford envisioned this city and factories to be the first of the "Hydroelectric Chemical Age." He argued that this would be an age of great industry but of clear skies and environmentally friendly operation. Ford said, "These factories will not belch forth smoke to the sky, nor anything like that. Although coke will be used in the fixation of atmospheric nitrogen [for fertilizer production], no coal will be necessary for heat, light, power, or smelting of ores."[12] Ford's Chemical-Electrical City would still need the coke and additional electrical power from coal, but his plan was truly visionary.

Ford had purchased extensive coal lands in Kentucky about 200 miles from Muscle

Shoals as alternative energy. Ford's plan was to maximize efficiency by burning coal at the Kentucky mines, generating power to send to Muscle Shoals versus shipping coal to Muscle Shoals, wasting transportation. He argued that sending electricity via the wires was far more efficient than shipping coal. He would, of course, have to ship the lighter coke for some industrial uses at Muscle Shoals. This additional power would be necessary in seasonal low-flow times on the river. Ford planned for huge power usage at Muscle Shoals with a huge aluminum plant. Aluminum production required cheap power and a good source of aluminum bauxite ore. Ford engineers had confirmed large deposits of bauxite in Alabama that could be used. Aluminum smelting needed vast amounts of electrical power and was only produced cheaply at Niagara Falls. Ford planned to increase aluminum in his Model T, reducing weight and making them more fuel efficient. Furthermore, Ford planned to make steel cheaper using electric furnaces at Muscle Shoals. He would also have more traditional blast furnaces running off the iron ore of nearby Alabama.

Electricity would power his huge seventy-mile city as well. There was to be electric passenger and freight service to Memphis, Nashville, Chattanooga, Atlanta, and Birmingham. Just as big was Ford's dream for the farmers of the Tennessee Valley. Ford planned to manufacture all three components of commercial fertilizer — nitrates, phosphates, and potash. Nitrates would be made from electricity and coke, phosphates would be quarried and crushed, and potash would be gained from Georgia shale and electricity. Ford estimated he could produce these at the lowest cost in the world. With available fertilizer, Ford even predicted the cotton growers would get a half a bale more per acre. It was a vision of industry and agriculture coming together in a unique way. For the vegetable farmer, Ford proposed, "Along the lake, on both sides, piers and boat houses will give many farmers opportunity to carry their vegetables, poultry, eggs, and dairy products by motorboat [electric] to the town and factory."[13]

Ford furthermore envisioned a complete harmony between rural and urban areas. He predicted lighter aluminum machinery making farming cheaper and more efficient. His factories would not only run on clean electricity, but waste and by-products would be applied and recycled. Ford said, "There will not be a dump heap at any factory. Everything will be utilized. That which becomes residue in one process will be made to serve a purpose in another."[14] To keep his electric metropolis's air clean, he proposed that all cars in the area run off grain alcohol made in the surrounding farms. Ford's vision gained massive support from the National Grange, the Farmer's Union, and the Federal Farm Federation. Even the American Federation of Labor (AFL) supported Ford.

Ford garnished strong support from the black cotton farmers of the South. Small black cotton farmers lacked money for the fertilizer and insecticides needed for cotton. Nitrates were badly needed in southern soils. This political push by Ford would be his first cooperative effort with Tuskegee Institute. R. R. Moton, the president of Tuskegee Institute and Carver's boss, spoke on behalf of the Ford proposal before Congress. Moton said, "If Ford gets Muscle Shoals, he will make calcium arsenate to kill the boll weevil, whereas today we cannot get arsenate in sufficient quantities to prove effective at any price."[15] Ford helped pay for the circulation of Tuskegee Institute bulletins on the need for fertilizer and how to control the boll weevil. Ford found immediate support from Alabama and most southern congressmen. Alabama, Oklahoma, and Nebraska state legislatures passed resolutions supporting Ford. Scores of chambers of commerce from midwestern and southern cities sent letters of support.

For several years, Muscle Shoals was the main focus of Ford. He was fully committed to his Muscle Shoals project, and he told a reporter in 1922 he planned no more auto factories or new cars.[16] Congress and the bureaucracy would nix Ford's grand plan for Muscle Shoals. He initially had the support of Secretary of Agriculture Henry A. Wallace (friend and student of G. W. Carver), but the political opposition grew. A push back from Washington created a massive show of support for Ford. A national organization called "Give Henry Ford an Opportunity Club" solicited one- to twenty-five-dollar donations. One mass rally in Mobile, Alabama, drew over 5,000 supporters. Ford had strong political enemies, and the progressive movement for more government ownership in power projects was at its peak.

The opposition became formidable as enemies united. Ford's own opinions always hurt in the political arena. His anti–Semitic views had made the papers across the country at this point. His support of black workers had made him unpopular in the South as well. The idea of improving the economic lot in the Muscle Shoals area was not popular with all the white residents. Another powerful lobby against Ford was the eastern banks. From 1921 to 1922, Henry Ford had stunned the banks into refinancing from within, reportedly saying "he would tear down his factories brick by brick with his own hands before he would deal with Wall Street."[17] Furthermore, Ford promised the farmers he would keep Wall Street out of the deal. New York's J. P. Morgan financed the opposing Chemical Trust. Still, for the farmers, the most hated enemy of the 1921–1922 farm recession was the eastern banks.

Ford found opposition from some of the regional power companies as well who feared Ford's entrance into electricity generation. Figures suggested Ford had the potential with Muscle Shoals to undersell most southern power companies. The "fertilizer trust" feared that Ford, with Edison's help, would drive the cost of fertilizer down and them out of business. Rumors fueled a real estate boom in the area, and lots were being brought up and even sidewalks were being built. The opposition in Congress was led by liberal senator George Norris, chair of the Agriculture and Forestry Committee, who wanted government ownership of Muscle Shoals. Norris was successful in blocking Ford, and ten years later having the government take over the dams under the Tennessee Valley Association.

The proposal blocked in 1923 came up again in 1924 when Ford talked about entering the Michigan Senate race. Ford had even been considered for the presidency. In the magazine *Collier's*, Ford showed a substantial lead over other candidates such as Warren Harding (who would win), Herbert Hoover, William Jennings Bryan, and Robert La Follette. The issue of Ford's project continued back and forth, but Senator Norris of Nebraska never wavered in his distrust of Ford and "Big Business." Norris went against both legislative and grassroots support for Ford in his state of Nebraska. Norris was the leader of the national progressive movement and would be one of Franklin Roosevelt's strongest supporters of New Deal farm policies. The legacy of Muscle Shoals is sometimes known as the "Alabama Ghost" of the Senate. It was the beginning of decades of battle over privatization versus government ownership. Farmers and locals of Muscle Shoals were extremely disappointed. While the government successfully brought cheap power to the South through the TVA hydroelectric dams in the mid–1930s, the other industries of the Ford proposal never came to be. The farmers surrounding the area remained some of the nation's poorest people. Ford's aluminum-body car never materialized because of a lack of cheap power to make aluminum competitive with steel. The nation continued to import large quantities of fertilizer for decades. Today Muscle Shoals is a small town rather than the South's largest city and the fertilizer capital of the world. Ford would move on to other utopian possibilities in South America and Southern Michigan.

10

The Sages of Dearborn and Tuskegee

"Mankind passes from the old to new over a human bridge formed by those who labor in three principal arts — Agriculture, Manufacturing, and transportation."
— Inscription on Ford Engineering Laboratory in Dearborn

Henry Ford and George Washington Carver were dreamers and doers. They both could be controversial, having strong beliefs. Ford was the more problematic with his anti-Semitism and hatred of bankers. Carver's strong faith caused him problems in the scientific community, and his moderate approach to facing racism caused him problems with black activists. They both believed in a better America through self-reliance. They wanted to change how Americans think. Carver had the more difficult path in his ability to change things. He lacked Ford's money and lived in a South of racism and separation. Still, Carver did leave his mark on improving the South socially and economically.

Carver's work in the late 1920s continued along similar lines of finding new crop uses and developing potential industries for the South, but culturally he was also changing minds. His fame had spread internationally, and he had invitations from Russia in 1930 to come and educate farmers, and a student of his did make the trip. He studied the geology of Alabama with the same interest in economic development. He often talked on the quality of Alabama marble and its extensive metal ore deposits. While he had failed in his personal efforts to commercialize the vast and varied clays of the area, he continued to look for new applications. The artist in him drew him to the beautiful colors of Alabama's weathered granite clays. He set up his own pottery shop to explore this industry. He found that these clays could not only produce colored stoneware, but fine china. He wrote to the Southern Newspaper Publishers Association in 1927 noting the potential: "Creative and applied science is coming to our relief as never before, and showing us the fabulous wealth we have in our own varied deposits of clay ranging in color from snow white kaolins and china clays, to the choicest of ochers, rare ambers, Vandyke browns, choice Indian reds, beautiful siennas and rare deposits of Fuller's earth."[1] With success in materials engineering moving along, he turned to other fields.

Not surprisingly, Carver's search for new uses of peanuts and sweet potatoes would lead to medicinal uses. He had worked over the years with face creams and massage oils. He drew on his experience as a trainer in college for the football team as well. Vegetable oils appeared to have more restorative powers than mineral oils. He had early on been using peanut oil to help relieve the pain of railroad travel. While Carver often mentioned his medical recipes, he was never comfortable trying to sell them. Yet his interest in medicinal plants had been extensive. He had a real interest in diet and health but never pushed it as hard as his interest in commercial crops. He often ate a variety of wild plants. His wild salads of dandelion, alfalfa, peppergrass, watercress, and lesser-known plants such as shepherd's purse, lamb's quarters, and sheep sorrel required an adventurous palate. He did find a willing taker in Henry Ford when they both shared a "weed sandwich" for reporters.

Like most American small farmers, Carver was well aware of the many home remedies and folk cures. Wood tar creosote made from beech wood and pine was popular in a number of home and commercial remedies. It was a powerful antiseptic. For years many had suggested it for cholera. It had been used as a home remedy for coughs and treatment of wounds. It was also sold in commercial medicines as a treatment for tuberculosis and sore throats. Creosote had strong medical support, but it was not consistent and had subsided by the 1920s. Over the years Carver had used it for sore throats by mixing it with peanut oil to improve flavor and add some nutritional value. Carver had used the mixture personally and would give it to students. He believed in its curative properties and had mentioned it often when he gave talks. Carver called his mixture Penol, and the medical value of the mixture was spread by word of mouth.

Throughout the 1920s, Penol gained commercial interest based on creosote's folk remedy popularity and Carver's own personal popularity. Clearly, Penol was far from a medical breakthrough, but it had popular appeal. This type of product, however, offered high profits because of the availability in large quantities of both peanut oil and creosote. A small bottle offered huge returns but would require advertising capital. Carver was able to put together some local investors to produce Penol. In 1926, the Carver Penol Company was formed as part of his bigger move into commercial paints. Penol was copyrighted but never patented. Carver never fully lent his name to the advertising if problems should occur, probably not wanting to upset the peanut industry which was a major source of income. Penol was never a big success as there were quality issues and complaints with such a product. Still, it had a loyal following.

In the early 1930s, another company took over Carver's interests in the Penol Company. The new company was Herb Juice–Penol Company. Its owner set up a deal for Sharpe and Donme Laboratory to manufacture the product. Carver was brought in as a consultant on its manufacture. The agreement gave Carver a monthly check of $100 and 2.5 cents royalty per bottle.[2] Sales never really panned out and problems continued. Creosote medicines in general were no longer popular, as questions about its toxicity started appearing in the press. Carver and his investors were also cheated out of royalties as well. Efforts began to restart manufacture of a new and improved Penol. The final crushing blow came from the Federal Drug Administration which invalidated some of the advertising claims.

One product Carver was more comfortable with was pure peanut oil as massage oil and rubbing oil with restorative properties. It was far from miracle oil, but it did have vitamin E and other ingredients that made it superior to mineral oil. Carver was well aware of the historical healing uses of olive oil and believed peanut had even greater properties. Since his days as an athletic trainer, he had realized that oil massages could be helpful in injuries and had used it himself after hard railroad or wagon traveling. The product had been offered as a commercial product. Carver, however, had come to learn that the South was not the place for a commercial entrepreneur. The South was only interested in cotton, and getting financing for health products was next to impossible. The South was King Cotton and the Princes of Tobacco and Rice. Ford had also run up against King Cotton at Muscle Shoals. Henry Ford was running up against a wall for creative farming in the industrial North. So the late 1920s and early 1930s brought both men back to projects more consistent with their regions.

Ford would also try his hand at health care. One health project of Ford's was the Detroit Hospital that he had started in 1914. Ford took over the hospital as the City of Detroit slowed in making progress. As with all of Ford's projects, his hospital proved revolutionary and controversial. It was a "closed" hospital, which charged very low fixed fees not based on a patient's income. No poor person could be turned away. It had a total of 486 rooms and 120 doctors. The doctors were on salary, so the hospital serviced the poorest of the city. The American Medical Association argued that doctors should be paid on fees based on the patient's income, which favored a preference towards the higher-income patients. The hospital serviced much of the homeless of Detroit and even offered temporary housing. Ford offered free medical care to children in need in Michigan. He combined his hospital with his interest in America's diet. Ford brought his boyhood friend, Dr. Edsel Ruddiman, from dean of Vanderbilt University's Pharmacy School to his hospital to study the American diet. Ford also pushed his vegetable diet and soybean products at the hospital. He often visited the hospital to give patients his own views on diet and health. It was at the hospital that soybean milk was made for patients. Still, the hospital was too small a project to keep Ford busy. He looked to expand into all phases of health care.

Ford was building a complete health system by the late 1920s, and developing his own brand of paternal capitalism. Henry and Clara started a school of nursing in 1925. The school started graduating seventy nurses a year, and the Fords helped many financially. Ford also operated a convalescent school to help children recover. The school offered nature study, a library, woodworking, and endless entertainment. In particular, Ford emphasized nature study and bird-watching for convalescing children. Families or parents could also stay in rooms at the Convalescent School. Ford expanded his hospital to handle employee families and children from every Ford operation. By 1930, he had invested over 6 million dollars in his hospital and health care system. This is a sometimes-overlooked project of Ford's, which gave much back to the community. The compassion of his health system continues as an American standard today. The system grew into the Henry Ford Health System, which today operates six hospitals and thirty clinics throughout Michigan. While Ford was doing much pioneering in health care, he turned most of the effort over to Clara in the early 1930s.

With the defeat of the Muscle Shoals project, Ford moved on to several other grand projects such as his South American utopia for rubber production and the completion of his industrial museum in Dearborn. Ford's museum had, in many ways, been a lifelong project and an enduring vision of industry and agriculture. He had started to collect farm equipment even before the first successful test of his Quadricycle. The project museum

(Edison Institute) and the outside museum (Greenfield Village) came together slowly, evolving out of Ford's collecting, his interest in education, and his vision of America. This project could also embody his belief in the fusion of agriculture and industry. It included fields of test crops of hemp and soybeans. At his Dearborn Farm he had all sorts of farm machinery, and at a nearby barn he had pieces of Americana from which he and his wife selected for their two restored inns. In 1927, Ford started to move things to his engineering laboratory in Dearborn, where he owned many additional acres. In 1926, Ford had purchased the surrounding land with an idea of a future museum.

According to Ford hagiography, Ford conceived the idea of an American history museum in 1919 on returning from a humiliating appearance on the witness stand in a libel suit. Ford had been suing the *Chicago Tribune* for an editorial that called him, among other things, an "ignorant idealist" and an anarchist. He was suing for $5 million. When Ford took the stand, the defense was determined to show his ignorance. The lawyers focused on American history, turning it into an oral exam. Unprepared, Ford was destroyed and made the fool. The news had a field day supporting one of their own and making Ford the fool. The jury found the paper guilty of libel but awarded Ford a mere six cents. Small-town America realized the attack on Ford for what it was. Typical of the responses was that of a small paper in Nebraska: "A few less smart-aleck attorneys and a few more Henry Fords, and the world would have less troubles and more to eat."[3] Ford was able to fuse this humiliation into fuel to drive his new American history. This type of positive "substitution" for revenge was typical of both Ford and Carver in personal attacks on them. Positive substitution is the hallmark of achievers and great people such as Ford and Carver.

As early as 1924, Ford had assembled a team of acquisition agents, drawing from the Ford Motor Company organization to build his outdoor museum. He added some professional antique dealers as well. Ford retained final approval on the purchases. He himself actively sought out antiques. He added mountains of farm machinery to the fields surrounding his Dearborn Engineering Building. Offices started to be filled with antiques. He engaged his engineering labs in the process of restoration. Ford tended to collect more ordinary examples of American life versus antiques. The collection lacked specific themes other than Americana. When the collection would be assembled for display, a reporter called it "the world's biggest rummage sale." As offices, closets, barns, and fields filled up with stuff, Ford moved up his plans for an indoor museum building.

The memory of the beating on history he took at the *Chicago Tribune* trial may well have played into his decision to make the museum an exact replica of Independence Hall. Ford hired Detroit architect Robert O. Derrick to design it. Ford demanded exact measurements and a perfect copy including any original mistakes. He not only had an exact copy of the Liberty Bell made at the original foundry, but the brick, soapstone, and granite used were from the original quarries. Where possible, all construction components were made by the original process. He even hired craftsmen in the older techniques where needed. In many ways, it was an engineering project on the grand scale of his Model T.

Ford began his new project of Greenfield Village by moving Edison's Fort Myers, Florida, house and New Jersey laboratories to the land next to his Dearborn Engineering Building. Ford had an army of workers number each piece and location of every bottle to have it exactly rebuilt in Dearborn. Photographs were taken as well. Ford required a foot of Florida soil be taken for each square foot of the laboratory. When the lab was reassembled in Dearborn, he brought Edison to Dearborn to formally launch the project. On September 27, 1928, with the opening of the Edison lab in what would become Greenfield Village, Ford laid the cornerstone for his future museum — Edison Institute. That day Edison put a spade of their dead friend Luther Burbank into the wet cement. It was to symbolize the union of agriculture and industry. Edison would leave his footprints, signature, and date in the block that would be the heart of the Edison Institute Building and would grow to over eight acres. Ford and Edison would then start the steam engine to run Edison's laboratory around which the 200-acre outdoor Greenfield Village would rise. The land itself was a beautiful oxbow of the meandering Rouge River. It had been a favorite birding site for Ford as well. It was also a short distance from Fair Lane, allowing for daily visits by Ford. It would be a sanctuary for an aging Ford of childhood and career happiness.

The full vision of the village and museum was still not fully crystallized. Ford hired an artist, Edward Cutler, in 1926 to start drawing sketches for the future museum. Ford loved the concept of Colonial Williamsburg, but he wanted a village free of a specific region or culture. In fact, he wanted it to represent the full evolution of American industry and agriculture. Ford wanted to be educational from the start, planning to incorporate schools by some means. Ford also saw publicity and promotional value in the project. He had seen H. J. Heinz's success with his Atlantic City pier. It also offered a project that both he and Clara could pursue. Clara had always enjoyed the restoring of the two inns they had purchased. She began to collect American pewter and silver pieces for the future museum. Ford let the museum and village grow, twist, and evolve to fit his ever-changing mission. He used his car dealers across the nation to hunt down antiques and artifacts. It become personal too, as the village was modeled after the village of his wife that had been swallowed up by the city of Dearborn.

The year 1929 would be one of passion for Ford's new project, but he mixed in travel as well. Edison and Ford would host President-elect Herbert Hoover at their summer residences in Florida. Later in the spring, Henry and Clara visited the restored colonial town of Williamsburg, Virginia, where Ford drew inspiration from for his Greenfield Village. In the summer, the Fords and Edison visited the Chautauqua Institution in New York which was an educational movement started by Edison's father-in-law, Lewis Miller. Chautauqua fairs brought learning to middle America. The Fords had been active in the Chautauqua movement in Detroit prior to their marriage. Interestingly, George Washington Carver was on the Chautauqua circuit in the South as a speaker. Ford took notes on things he could incorporate in Greenfield Village, and the village is truly a mix of many historical themes.

Finally, in October of 1929, Edison and Ford traveled to West Orange, New Jersey, to look at moving Edison's original Menlo Park research center to Greenfield Village. Menlo Park had been the site of the discovery of Edison's best-known inventions such as the incandescent light and the phonograph. Most of the laboratory had been taken down, but this did not deter Ford from purchasing the original boards where he could. Ford went over all details of the restoration with Edison and his industrial archeologists. As always, Ford tended to produce his own experts from the Ford organization, limiting outside expertise. Ford wanted to be in full charge. He included the brick office, Edison's machine shop, and boarding house for movement to Greenfield. He even included twenty railcar loads of New Jersey soil. Ford would personally oversee this project for months.

The museum project started to consume Ford's time. His chief operating manager, Charles Sorensen, said, "After Ford started Greenfield Village and Museum at Dearborn, he seldom came to the Rouge Plant…. In later years he actually put more hard work into the Museum than he did into Ford Motor Company. The Museum was to be his monument — a cross section of American industry, beginning with early times, and a visible, authentic record with tools and machinery of all arts of manufacturing originating in the United States. He loved Greenfield Village."[4] While the museum and village looked at American history, it was also very personal, including the Ford family farm, the schoolhouse he attended, the home of his favorite teacher, the first Edison Electric plant he worked at, and many other places that not only emphasized American but Ford's personal history. What is truly amazing is that Ford didn't separate these projects in his mind but saw them as an integrated effort to educate America.

Another one of the earliest buildings in Greenfield Village was the Martha-Mary Chapel. This building represented a historic creation of many things and events in Ford's life. The building name was after Clara Ford's mother (Martha) and Ford's mother (Mary). It was the creation of Edward Cutler, Ford's architect. The home of Clara's parents where Henry and Clara had wed had been torn down in 1920. Ford was able to salvage 9,500 bricks, hardware, and timbers to use in the building of the chapel. Many unrelated antiques were used in the building such as the chapel bell which had been cast by the son of Paul Revere. Ford began to move mills and shops into the village, which could also play a role in making the materials needed for the village reproductions such as rugs and carpets. In 1929, the Martha-Mary chapel functioned as a school for the start of Ford's private school to be interwoven in the village. During the construction phases before the village opened, it was already getting 400 visitors a day. Ford would use the same model to build chapels throughout the United States.

While the village progressed, the museum moved quickly to completion in the fall of 1929. The goal became the opening date of October 21, 1929, the Golden Jubilee of Edison's incandescent light. In reality, it would take many more years to complete, but Ford had enough in place to plan a huge celebration. Part of that would be a $300,000 investment in a teak wood floor from the East Indies, so the museum would last a thousand years. Ford was able to get the furniture collection in place for the 300-person Jubilee Dinner. Many of the furniture pieces were antiques owned by early American patriots. A full musical instrument collection of colonial times was also in place. The balance of the museum was evolving in three sections based on Ford's principal arts of agriculture, manufacturing, and transportation. Ford's Greenfield Village was far from a traditional museum. He mixed a school system, experimental farm, and a green laboratory in the village.

Carver had cemented the title of the "Peanut Man" by the 1930s, and Ford was well on his way to becoming the "Soybean Man." Ford had been interested in soybeans for a long time and initially was attached to them for their food value. However, in 1929, Ford built his soybean experimental lab in Greenfield Village. He hired self-trained chemist Robert Boyer to lead a team of Henry Ford Trade School graduates to look at the industrial uses of soybeans and other plants. Actually, Robert Boyer was the son of Ford's manager of his restored Wayside Inn in Massachusetts. Ford set up another soybean lab to look at the food value of soybeans. Boyhood schoolmate Dr. Edsel Ruddiman, who had progressed to the head of the pharmacy school at Vanderbilt University and came to Ford's hospital in 1926, headed this lab. At the Henry Ford Hospital, Ruddiman had started work on a "vitamin biscuit" for Ford. At the new soybean lab, Ford had directed Ruddiman experiments with wheat, soybeans, and other vegetables to eliminate the need of the cow to produce milk. In the 1930s, Ford's major interest, drive, and passion centered on soybeans. He saw soybeans as the keystone between agriculture and industry. In fact, Ford, Ruddiman, and Boyer had spent the first years at the Greenfield lab searching for the vegetable with the most overall potential. The result was the soybean. From 1930 to 1932, Ford spent over a million dollars in soybean research. He proudly held a National Convention of the American Soybean Association in Dearborn in 1931.

By 1932, Ford had planted twenty-five acres at Greenfield Village in over 500 varieties of soybeans. His extensive Michigan farm network had over 78,000 acres supplying him with soybeans. He set up soybean oil processing plants at the Rouge River plant and the village industries at Milan and Saline. Ford was producing 500 tons of oil for use in car paints. These plants, in addition, were producing over 100 tons of oil, which was used as fluid in shock absorbers. Ford was replacing linseed oil in all his paints at a savings of over 35 percent of the cost of linseed oil, using about a half gallon of soybean oil to paint a car and another half gallon for the car's shock absorbers. The remaining soybean cake was used in plastics and food items. At the Rouge, soybean plastics were used in distributor cases and electrical device covers. Ford was even mixing and pressing cellulose fiber and soybean cake into tractor seats. Ford used another 200,000 gallons of soybean oil for binder in his cast-iron sand molds in the foundries. Ford was particularly proud of his soybean ice cream and flour as well. The flour was being used at the Ford bakery for employee cafeterias and grocery stores. Ford could not supply his internal needs for soybean oil, so he organized his extensive farms in Southeastern Michigan to plant soybeans and did the same for his Cooperative Farmers Association.

Ford's network of farms had started back in the 1920s when he used the profits from Model T production to buy farmland in Michigan. Ford had a type of control over the farmers in his cooperative based on his financial help. He had freed them of the banks in their mortgages. They could purchase gas and oil at one cent a gallon versus the market price of five cents. Over 400 Fordson tractors were available for cooperative use. The farmers got reduced repair costs for their tractors and cars. Ford offered low-cost milling and storage of their grains as well. In addition, Ford ran a commissary for the farmers. In all, Ford was able to use his network of 700 farmers to put 22,588 acres into soybean production; and in addition, Ford contracted for an additional 15,000 acres in soybeans. Ford's price of fifty cents a bushel made soybean production attractive during the Depression. Even today this area of Michigan remains a major producer of soybeans. Ford engineers designed a machine for soybean oil manufacture at his Saline plant. He used gasoline solvents and a distillation method. Even so, Ford could not meet his needs for soybean oil in the processing of automobile paint.

Ford was always a farmer buying and expanding acreage every year. Ford owned America's largest farm with acreage amounting to over three million acres or roughly the size of the state of Connecticut. He grew every plant imaginable. Ford set up many roadside stands to sell his huge vegetable crop in the Midwest. He hired thousands of workers and school boys in the fall to harvest his crops at good wages. Ford's farm employment rivaled that of the federal government during the Great Depression. The majority of his farms showed a loss; in fact, in the 1940s the IRS refused to believe Ford was producing a million-dollar loss every year.[5] This was Ford's biggest love and hobby.

One of Ford's proudest moments came with his exhibit at the Chicago World's Fair of 1933. The exhibit was called the "Farm of the Future." Ford built a mini-processing plant for soybeans and a few acres of soybean crops. The exhibit included an "industrialized barn," which included a reconstruction of his father's barn in Dearborn. There were cases showing not only soybean plastic car parts, but radio cabinets, glue, various enamel paints, soaps, margarine, and others. Ford was also proud of his soybean plastic/paper tractor seat. Ford served the following dinner for guests at the fair in 1934:

> *Tomato juice seasoned with soybean sauce*
> *Salted soybeans   Celery stuffed with soybean cheese*
> *Puree of soybean   Soybean crackers*
> *Soybean croquettes with tomato sauce*
> *Buttered green soybeans   Pineapple rings with soybean cheese*
> *Soybean bread and soybean butter*
> *Apple pie (soybean crust), soybean cakes, soybean cookies, soybean candy*
> *Cocoa with soybean milk   Soybean coffee*

More than 25 million people would see the Ford exhibit. Deep-fired soybeans were handed out to visitors as a snack. Ford produced a nineteen-page recipe book on soybeans to pass out to fairgoers and workers. Visitors were offered a refreshing drink of soybean milk. Ford himself became a lover of refrigerated soybean milk, often flavoring it with maple spray or honey. By 1942, Ford's Dearborn laboratory was producing 150 gallons of soybean milk a day for distribution to his hospital, lunch rooms, grocery stores, schools, and for delivery to friends.

Ford's soybean milk had the same composition as cow's milk: 3.5 percent protein, 4.8 percent carbohydrates, 3.5 percent fat, and .7 percent minerals. Doctors at the Henry Ford Hospital had enabled rats to live six generations using nothing but soybean milk. The dairy lobby started to take notice, and laws were passed to prevent sales of items like soybean whipping cream. One dairy product that Ford scientists failed to perfect was soy cheese, which lacked the texture and flavor of milk cheese. Still, in 1934 Ford had defined a new future in vegetable-based dairy substitutes.

11

Green Supply Chain for America's Industry — Village Industries

> "What is past is useful only as it suggests ways and means for progress."
> — Henry Ford

Like Ford, Carver believed in a decentralized approach to agriculture and industry. In particular, both men believed food should be grown close to the point of consumption, and similarly manufacturing and processing should be at the local market level. In 1909, Carver noted the economic inconsistency of a local farmer not having a garden: "I took dinner in a country home, this was the bill of fare, flour from Minnesota, coffee from Brazil, macaroni from Italy, cheese from Wisconsin, bacon from Kansas, and cake made from eggs purchased at the store. In passing through the country I find a number of so-called farmers without a garden of any description."[1] Ford saw the same inefficiencies of the population in cities dependent on faraway farms. Both Ford and Carver would approach the problem but from different ends. Ford planned to use the same community-owned, efficient, and clean manufacturing throughout his whole industrial supply chain. Carver believed in a self-sufficient farm.

Ford had been building manufacturing communities since his design of the River Rouge plant. Even then he was starting to consider a supply chain of village factories that would bring farm and factory together. He hoped to tie agriculture and manufacturing to each other in physical and environmental harmony. Ford had first tested the concept of community manufacture with the building of his Ireland plant. This plant had lifted an economically depressed area of his family's roots into a striving community. He built a supply chain of local industries to support his manufacturing plant. Ford was always proud to point to the economic turnaround and uplift of the society of Cork.

Ford had looked to further expand community manufacturing along his assembly plant supply chains. His lumbering operations in Northern Michigan was one of his first community manufacturing efforts. Here he built whole towns to support his workers. Ford loved this type of social engineering and environmental control. The isolation of the lumbering operations allowed him to impose his image of community. Ford's Iron Mountain community was maintained by using hydroelectric power from a dam he built. Furthermore, Ford imposed his environmentally friendly practices in lumbering. He did impose codes of behavior prohibiting smoking while working and preventing town commissaries from selling tobacco and alcohol. Ford saw these two items as ruinous to humans. Ford repaved streets and built schools. Carver agreed totally with Ford on tobacco and alcohol. He had argued in his bulletin, *Some Ways to Save Money on the Farm*, of the total waste of alcohol and tobacco. And like Ford, Carver demanded that homes and buildings of the poor southern farmers be painted and cleaned often. Also, as in one of Carver's bulletins to farmers, Ford argued for the planting of flowers around worker houses. Campsites were to be neat and clean. Ford, like Carver in farming, found that cleanliness and neatness were keys to morale. In the 1940s, the Japanese studying Ford's practices came to the same conclusion about cleanliness.

Ford's community manufacturing successes led him into other areas of the supply chain. After an interruption of coal supplies in the early 1920s, Ford purchased mines and a railroad to deliver coal to his factories. Ford looked to apply community manufacturing concepts to one of the nation's dirtiest industries. He believed in the same green and environmental dreams for the coal industry. In the 1920s, he had purchased coal mines in West Virginia and Kentucky to assure an uninterrupted supply. Ford cleaned up the mines and the mining town. He would describe the project: "First of all, we cleaned up the mines and their surroundings—a mine can be clean. Such of the houses as were not worth painting, we replaced with good houses that had bathrooms, we put down sidewalks and hard surface roads, we put in street lights and a recreation building, and tried in every manner that we knew to make the little towns into first-class places in which to live. We put in our regular wage scale, and the men are now earning about twice as much as other miners in the field."[2] Ford's ultimate dream was to produce electricity at the site of the mines, saving transportation costs.

Ford's automotive supply chain offered endless opportunity for village industries. Ford's American factories were huge consumers of cotton, steel, glass, rubber, wood, coal, paint, and a massive array of subassemblies and manufactured parts. Nature and man had often threatened his huge supply chain. The boll weevil had for years threatened Ford's need for quality cotton for his cars and tires, driving him to look into growing cotton. Strikes had threatened his supplies of iron ore and coal, forcing him to buy mines and railroads. For the most part, his massive supply chain of raw materials was domestic, with the major exception of rubber. Rubber plants were grown in South America, Africa, and Asia, and then shipped to the auction markets in London. England's control of the wholesale market gave it a cartel or monopoly over world rubber. The semi-raw rubber was then shipped from the London markets to the rubber factories in Akron, Ohio. Such a long and international supply chain increased costs and came under many international threats. Ford saw this as an ideal possibility to expand his green supply chain.

Ford and Harvey Firestone had early on recognized the potential problems of non-domestic supplies of rubber in the buildup to World War I. Ford, Firestone, and Edison had been doing extensive research into a replacement of rubber plants. The search for domestic rubber had proved difficult for even the best scientists such as Thomas Edison. Ford started to look at an improved supply chain solution. South America, where the rubber production had originated, had lost its monopoly to Asia. Firestone had discussed the possibility of buying plantations in South America with Ford. In 1925, Brazil's consul,

Custodio Alves de Lima, visited Henry Ford in Dearborn. Ford already had auto factories in Brazil and produced 60 percent of the cars there. Ford trucks and tractors were popular as well. Brazil was interested in bringing the Ford system and methodology to its declining rubber market. Brazil, too, was suffering from the British cartel that favored the Asian rubber plantations. In 1913, an agent of the British rubber cartel had smuggled out 70,000 rubber plant seeds and planted them in British Asian plantations. Brazil controlled 95 percent of the rubber in 1913; by 1925, it was under 5 percent.

Brazil was willing to offer Ford many enticements such as tax breaks. The American government also wanted to break the British cartel and offered some help as well. For Ford it was another opportunity to apply his lean system of organization and social engineering. This was a different experiment in agriculture and industry for Ford. At Fordlandia, Henry Ford tried to bring mass production principles to agriculture. Ford sent botanists to the Amazon to further explore the possibilities of rubber growing. He would underestimate the problems of the tropical climate, South American politics, and its culture. Ford, however, was confident that his system had no ideological or international boundaries. In 1927, Ford was given 2.5 million acres of land on the Tapajos River to establish a rubber plantation. The plan was to produce enough rubber for 2 million cars a year. Ford would pay the Brazilian government 7 percent of the profits and the local government 2 percent. Of course, Ford's vision was of a city on the hill in the heart of South America.

He envisioned a new type of plantation based on his production system. Ford saw the potential for a worker city that had never evolved in America. In this case, Ford would have to bring the workers to a new area and build a community from the ground up. Ford wanted to maintain the high wage system he had pioneered in the United States, but this workforce would be much different. It was more than pay; Ford would have to supply housing, food, family services, community services, and medical care. The plantation system was a necessity, but Ford did not want the slavery of the plantation system. It would be a hard lesson for Ford on South American culture, but it would later spur Ford's historical interest in the old Southern plantation near his land at Ways, Georgia.

Ford's idea of an industrial plantation based on his manufacturing system in Brazil was a doomed project. The workers had little use for excess money and no understanding of time. They were subsistence hunters and gatherers at heart. Ford engineers and managers had little means to manage or motivate them. Ford gave them much in terms of hospitals, schools, and houses, but in their need they saw themselves more as slaves than workers. Ford's regimentation and enforced routine, while forced in Detroit, felt like slavery in Brazil. Fordlandia in its first two years saw riots, fights, and strikes. Ford was not implementing a new manufacturing system but was trying to change thousands of years of culture.

By 1930, Ford managers realized that this rubber plantation was a bridge too far for the Ford system, but Ford was far from ready to give in. He had been studying the early rice plantations near Ways, Georgia. They had practiced a more humane type of plantation system known as the task system. The task system gave the slave something to work for, and they were given land to build on and work. Once the tasks of the plantation were

Cottage House at Ford's rubber plantation in Brazil, 1928 (from the Collections of the Henry Ford, Benson Ford Research Center, Photograph THF95643).

complete, they were free to do things and plant their own food. Interestingly, George Washington Carver argued for a type of task system to be applied to the economic slaves of the tenant farmers of the South who had been forced to plant all their land in cotton. The tenant could not even plant a simple garden because it took away from cotton production needed for rent payments. Carver argued that these tenant farmers would be more productive if they had some land to feed themselves and profit a little from a garden. Variations of the task system were as old as America herself. It was the use of private plots of land that saved the pilgrims from the economic failure of communal farming.

Ford put together another approach in his South American rubber plantation that was very similar to the old task system, which gave workers housing and one-quarter acre to garden. He added recreation and tried to create commerce for the workers' money. Still, Ford and his managers missed the bigger point in always feeling that Brazilians were just Americans waiting to be converted to the American way. Houses, recreation, and even medical care were based on the American way and designed in Dearborn. The other major problem was that rubber tree plantations were not the same as an assembly. Ford's vision of "Rubber Rouge" was not possible. To a large degree, this was God's territory, not that of machines and human routine.

Fordlandia had varied too far from Ford's vision of harmony with agriculture and industry. Rubber plantations in South America were not industrialized farms. The failure of Fordlandia, however, was not so much, as critics suggested, a failure of "Fordism" or the Ford process methodology, as a cultural problem. Ford's industrial efficiency depended on saving time. The Brazilians had no concept of time, yet they waited in line to punch a "time card." They worked eight hours, of which they had no concept. They were regimented and followed routine; they were given food, but it was the food of a strange culture. Their hot new Cape Cod–style houses seemed like prisons compared to the airy huts where they grew up. Their supervisors and bosses were of a different race and spoke a strange language. The medical aid was painful and disturbing. It was no wonder the Brazilians felt like slaves or even "dogs."

Fordlandia was too far away from both Dearborn and the Brazilian government. Henry Ford returned to his original vision of bringing agriculture and industry together in the American setting he understood. He always believed in factories mixing in harmony with surrounding farms in Michigan. He was never fully comfortable with the size of his massive River Rouge integrated assembly. The River Rouge plant had evolved out of Ford's wanting to fully own the supply chain (vertical integration) just as the assembly line had evolved out of his quest for faster production. River Rouge had started as a meta-village of manufacturing plants such as an engine foundry, plastics department, carpenters shop, and so forth to feed the Highland Park assembly plant. Massive assembly lines had been more the plan of Ford's managers such as Charlie Sorensen to meet Ford's overall requirements.

Ford had early on envisioned a chain of village plants powered by water on Michigan's rivers (Rouge, Huron, and Raisin) supplying centralized assembly operations. He hoped to use the rivers for power generation and man the factories with part-time farmers. Ford wanted to avoid the clutter of big cities and the loss of natural woodland. These village factories would be in harmony with their rural surroundings and also have large gardens. He hoped that rural values could be developed and maintained in such an environment. There would be less need for government and little crime. In this respect, he didn't need to change American culture but only merge the segments of agriculture and industry. This is what he had envisioned in Muscle Shoals, a seventy-five-mile string of industrial communities in a rural corridor. This was a vision unique to Henry Ford, which had been his dream from childhood. Ford believed that these connected village industries could also level out the boom-and-bust nature of big capitalism. These village corridors would be self-sufficient and have a Jeffersonian agrarian independence.

Ford's interest in village industries went back to the early 1920s. It would grow in 1944 to a total of nineteen village plants. The Northville, Michigan, plant had proven to be highly profitable in the manufacture of engine valves in the early 1920s. Some academic historians have treated Ford's village industries as a Ford hobby, a union-busting strategy, a revival of communal manufacturing, and even a neo–Marxian strategy.[3] Some argue it was part of the general movement in American business to decentralize. While Ford talked of decentralization for his village industries, it was not the force for the village industries. The reality was that this was a deep-seated Ford vision of agriculture and industry that had been evolving since his youth. These village factories were fully consistent with Ford's view of industry and agriculture. The village plants were built with a disregard for cost because Ford saw them in a bigger picture. He did often use these village factories for industrial and social experimentation. In his 1926 book, *Today and Tomorrow*, Ford stated his motivation for the village industries as being "to get rid of the overhead of the big city [taxes and land costs], to try to find balance between industry and agriculture, and more widely distribute purchasing power of wages we pay among people who buy our products."[4]

The village industries were Ford moving from automotive engineering to social engineering. He even tested biblical applications at several plants, which may have resulted from his close relationship with Carver and Christian ministers at the time. At his Tecumseh/Hayden Mills plant, he embarked in the highly publicized "Dynamic Kernels" project. This project was based on the Quaker miller, Perry Hayden, who sold his mill to Ford to process soybeans. Hayden's project was Bible based on a tithe of 10 percent. In 1940, a cubic inch of 360 kernels of wheat was planted. The goal was to tithe 10 percent for the next six years. Ford purchased the acres needed in the future as it grew to 330 acres in 1945, producing 4,868 bushels. In its final year, it reaped a crop worth $100,000.

Some considered these village factories a "hobby" because Ford did not apply the rigid profitability accounting that he did at the Rouge and other assembly plants. His managers often complained to him about the lack of cost control, but Ford would have none of it in his village plants. Accounting of these village industries showed large losses, but in fairness, Ford included the costs of schools, churches, and the like. The profitability of these village industries is difficult to assess based on the company's own records. It seems that two sets of books were kept—one by Henry and his accountants and another by his son and his accountants. There is no doubt that while Henry was alive, the massive ball of wax known as Ford Motor made huge profits because it had synergy and direction. He focused that corporate energy, and it would not be sustained by the decentralized organization that followed his death. A 1933 article in *Fortune* magazine noted, "having constructed the largest automobile plant in the world at River Rouge, Mr. Ford had learned all centralization could teach."[5] In later years, the village industries were the only factories he remained personally involved with. Shortly after Ford's retirement in 1945, Henry Ford II and his managers started shutting down these plants based on cost accounting analysis.

Henry Ford was willing to spend money on his agrarian manufacturing vision. The village factories were well kept, clean, landscaped, air-conditioned, and painted. Ford worked with the local communities, but these towns were not the company mill towns of Pennsylvania or the company mining towns of West Virginia. These were farming communities first, with a Ford factory. Ford did not try to impose the social norms of his city factories and his notorious sociological departments. He had learned that education was the key to cultural reformation in America. Ford did take an active role in supporting and improving community schools. With these schools, Ford imposed his McGuffey one-room schoolhouse approach. Scholarships for higher education were offered to the village communities. Ford also helped support churches. In community projects such as cleaning local rivers, Ford led the way in his village industries versus the "mandate" of his larger operations. Similarly, Ford promoted gardening and flower beds through prizes and awards.

Ford required the village factories to purchase supplies locally where possible, which was usually a boom for small-town hardware stories. Ford had seen the turnaround in Cork County, Ireland, through factory/community building. He developed an economic theory that such village industries would better distribute the wealth. This would, in the long run, make rural communities independent of large banks. Ford also believed this influx of money would bring automation to rural farms through their ability to purchase tractors and cars. Ford set up his village industries to be truly out of the community, setting rules to prevent city workers from commuting to his rural factories. Ford also set rules to distribute money into the hands of the unemployed, the handicapped, and single women.

The employees of the Ford village industries were made up from the surrounding area, usually restricted to about seven or ten miles. Part-time farmers were common and even preferred. Ford also employed women in his village industries (he had no women at his large assembly plants) to tackle more difficult assembly. Women were usually unmarried, unless their husbands were unemployed. Ford used handicapped employees from the community where possible. While women were paid less than men (this was the norm of the time), all Ford employees were higher paid compared to their counterparts in other industries or companies. Ford was paying a six-dollar-a-day wage in the 1930s. Few blacks or Mexicans were employed at the village factories because of the lack of these minorities in rural Michigan. Employment at the various village industries ranged from 17 to 1,200 employees. Ford set an interesting standard to employment levels using the horsepower by waterpower. He equated one unit of horsepower with one employee.[6]

The village plants represented a strange mix of operations. Ford used his industrial architect, Albert Kahn, to refurnish and/or construct factories at the locations of various gristmills. Albert Kahn had designed Ford's hydroelectric manufacturing plant on Green Island, New York, which would be a model for the village industry project. They were also modeled to some degree after his first two mills—Northville (1920) and Nankin Mills (1921). Ford tried to use waterpower but often needed other energy backup because of low flow on the Michigan rivers. The largest of the village industries was Ypsilanti, Michigan, with 1,500 employees, which made auto starters and generators. The smallest was Sharon Mills at 17 employees, which made cigarette lighters. Waterford (1925) employed the most skilled workers who made precision inspection gauges. Ford purchased the Swedish Gauge Company of Sweden and had the skilled machinists brought to Michigan. The Waterford workforce was not related to farming as the other village workforces. The Waterford employees were paid ten dollars a day, the best in Ford Motor's high wage system.

Ford had always looked to expand the use of disabled workers. He had a large number of disabled working at the Rouge earning standard wages. The Cherry Hill plant (1944) produced ignition locks and was manned by disabled veterans. Ford even built special housing for the disabled at Cherry Hill. To Ford's credit, he limited publicity on Cherry Hill so as not to put pressure on the disabled workers there.[7] Ford used Cherry Hill to test and expand the use of disabled and handicapped workers, including blind employees. Ford had always been a major employer of the handicapped, and at Cherry Hill, he proved their ability to work most jobs in the company.

The cost of these various factories ranged from $400,000 to $1 million. The Manchester plant (1936) was typical of the cost breakdown. The total cost was $831,000, of which $38,979 was for the eighty-eight acres. The major cost was $203,000 for the development of the lake and dam for power generation. Next was the building cost of $192,000 and the machinery at $192,000. The Manchester plant manufactured electrical gauges and was built on an old gristmill site. The plant had about 280 employees. Ford maintained from the 1920s through the 1930s that these plants were profitable. During World War II, several of the village industries were highlighted by the government for their efficiency. In 1922, Ford argued the success of Northville in having a cost advantage over Highland Park. Similarly, in 1926 Ford highlighted village production at Flat Rock in making vehicle lights. Ford's number was twenty-eight cents per unit cost at Flat Rock versus thirty-eight cents per unit at Highland Park. Still, Ford's major associates — Edsel Ford, Charles Sorensen, and Harry Bennett — saw these village industries as wasteful. The real numbers are not clear because cost accounting was often as much a political science as management science at Ford.

Henry Ford argued that the lack of overhead at these village industries made accounting more active and responsive:

> The bookkeeping and management of these plants is very simple. The records show how much material goes in, how much finished articles come out, and how many people are employed. That gives all we need to know. In the smaller plants, the manager attends to the records as part of his duties; while where more men are employed, the manager has an assistant who, in addition to other duties, keeps the records. None of the plants have offices or clerical staffs. There is no need for them — and that is a savings.[8]

Henry Ford asked for no distribution of overhead from Ford corporate, but he did include community payouts and projects. After Henry's death, cost accounting added corporate overhead, which impacted them negatively.

Ford's village factories and their communities had a special relationship as did Ford's village employees. Ford did not try to dominate as he did in Dearborn. He believed that these small communities were already idyllic utopias based on Midwestern values. Still, Ford loved to influence and cajole the communities to his way of thinking. He remodeled community schoolhouses to fit his McGuffey one-room model. He funded community gardens for the workers and community youth. He rebuilt churches and built community centers. The value system was important to Ford's experimentation at his village industries. Ford pressured workers to live within their communities and to maintain rural values and work ethic. These highly motivated employees allowed Ford to try new practices and methods that would have been impossible at his urban assembly plants.

One experiment was his management system. At his larger plants, Ford had a seven-layer hierarchy of management; but at his village plants, he used a three- to four-layer system based on the worker's better work ethic. At most, a village plant had a plant manager, assistant manager, general foreman, and foreman. Many of the plants had only a plant manager and foreman. Village plants had operational freedom, and young men were given more responsibility.

Ford even experimented with workers during their own inspection, a concept which was years ahead of its time. The ability of the worker self-inspecting was based on an extension of the crafts workmanship to the factory system. It required a very strong work ethic and a strong loyalty to the company's reputation. Like most American factories, Ford's larger plants had large inspection forces sometimes as high as 5 to 10 percent of the total workforce. Another difference of the village industries was their commitment to plant cleanliness. These factories were truly showplaces. Ford even experimented with a type of life employment. During slow times or model changeover shutdowns, employees were hired to work in maintenance, plant cleaning, and even in the community gardens. All these ideas are being once again looked at today in industry. The village industries of Ford should be required study for today's operations managers.

Ford built not only community but family at these village industries. There were family picnics and retirement parties. Employee birthdays were honored. Dances, company parties, and fiddling contests were common. To assure that plenty of country music was available, Ford created his own record label company. He also organized community sports. Ford believed that community building created peer pressure to do a better job. Ford organized sports competition between the village plants with the idea that it would facilitate communications about how to better run the plants. The village industries were truly experimental labs for employees and managers to try new ways of manufacturing. Ford even tried to run the village shops with part-time farm labor, but scheduling became a problem for a full conversion to such a system.

Ford picked up a lot of criticism of the village industries in the 1930s because the United Auto Workers (UAW) believed it was part of an antiunion strategy. The two things were clearly separate. Ford's antiunion strategy was very public; he opposed unionization. His reasoning was not about wages or benefits because he was paying the highest in the United States. For Ford, it was control; and he was unwilling to share decision making with the union. He, of course, had to allow the union in as the movement swept the industry in the late 1930s. The village industries were small and an insignificant part of the Ford Motor Company. They were not a major part of corporate manufacturing decentralization, which both Ford and General Motors planned to help slow unionization. Even management historians have tried to analyze the village industries as corporate decentralization, but his village industries were far more important to Ford than stopping unionization. Yes, it was decentralization, but it was Ford's unique experiment in agriculture and manufacturing that was driving it.

Did Ford believe that the huge automotive industries could be reduced to village factories? Probably not, but Ford was searching for some way of bringing the factory closer to the farm. Village industries could do that on a limited basis. Ford's strongest economic argument that was it would distribute wages to rural wage earners who could buy cars and tractors. Ford correctly foresaw that mega-integrated factories like the Rouge would not dominate the future. Communal manufacturing was not new, but Ford was the first to apply a modern industrial element to it. Henry Ford saw a communal supply chain, using the village industries or a regional mix of agriculture and industry as with Muscle Shoals. His son, Henry Ford II, and the next generation of Ford managers did not share Ford's vision. Henry Ford II and his associates started to close the village plants with the retirement of Ford in 1945. Henry became more and more isolated from the younger managers and his son in company affairs. Henry Ford, however, continued his industrial village experimentation and his cutting-edge creativity to the very day of his death. In fact, the very day he died, he had visited one of them.

Carver was just as active as Ford in this period. George Washington Carver remained creative as Ford but was slowing down a bit physically in the 1930s. He was fully doing research with no teaching duties, but he was in demand to speak and had a large volume of letters to answer every day. Carver had never been very good with assistants, but he was becoming a burden on the college's clerical resources. He was a typical Victorian, finding comfort alone in the laboratory. Carver did start to realize that physical restrictions were limiting his creativity. During the 1930s, Carver was regularly ill and had to be hospitalized often. In the fall of 1935, Carver added a young graduate, Austin Curtis from Cornell University, in chemistry. Curtis had several years' experience at North Carolina Agricultural and Technical College. Having an assistant was no longer an option; Carver needed help on many fronts such as research, letter writing, the museum, and the writing of his biography. Carver was slow to warm up to Curtis, but they grew close over the remaining years.

Curtis helped write Carver's biography, did research, answered letters, set up appointments, and traveled with Carver. Curtis was not of the same mold as Carver and Booker T. Washington. He was not tolerant of poor treatment of blacks and much more of an activist than Carver. He was young and an idealist and actually was a positive influence on Carver in standing firm on resisting poor treatment. Austin Curtis saw that Carver was not only an iconic figure of the blacks' passage from slavery, but a symbol for the intellectual equality of the future. One incident was in 1939 when Carver and Curtis traveled to New York for a radio show. Curtis had made the hotel reservations, but on arrival, they were told the hotel was "filled." Curtis would not stand down. He called Doubleday which was publishing Carver's biography. The publisher tried unsuccessfully to have the hotel honor its reservations. Curtis had representatives from the *New York Times*, *New York Post*, and the *Pittsburgh Courier* on hand. While waiting, a representative from Doubleday came and was offered a room immediately. Doubleday was now threatening legal action. With the hotel fully exposed and a corps of press writers, the hotel was forced to back down. Certainly, Carver in the past would have politely accepted the racism and moved on, but with Curtis, things would be different.

Curtis would play an important role in the laboratory as well. He continued Carver's work on paints, improving on both the color and sustainability. He worked on sources of more economical vegetable oils for soap making. In the 1930s, the United States was importing huge amounts of palm oil for soap making. The palm oil was very cheap, so peanut oil and most vegetable oils were not competitive. Curtis looked at continuing the possibility of Carver's idea of getting oil from magnolias. Later, Curtis would work with Ford on the potential use of soybean oil in soap making, which was cheaper than peanut oil. Chemurgists also picked up the quest to replace imported palm oil in soap making with a domestic alternative. Curtis also tackled some Carver-type projects of his own such as the making of artificial leather from pumpkin. Curtis's expansion on Carver's

11. Green Supply Chain for America's Industry—Village Industries

George Washington Carver (second from left) and Mr. Wickersham (far right) of the Georgia Railroad at a special car provided for Carver to travel to Atlanta (Bentley Historical Library, University of Michigan, File HS8578).

paints earned him a major government grant for painting low-cost houses at the Tennessee Valley Authority project.

Curtis once again looked at the potential for the commercialization of Carver's work. Carver had long ago given up on any commercial projects, but the late 1930s brought Carver back into the medical limelight. During the 1930s, America lived in fear of the crippling disease known as polio or infantile paralysis. The poor black South was being hit particularly hard by this disease that crippled and caused deformity from the wasting of muscle tissue if not treated extensively. With his compassion for poor blacks and scientific learning, Carver entered into the search to help polio victims. He approached it from his experience and background with muscle massage and peanut oil. Carver had been working with the children of some friends who had showed possible early stages of polio. One of these was from the prominent Thompson family that had commercial ties to Carver's early products. As Carver's methods showed some success, his friends leaked stories to the press, which was always on the lookout for a potential breakthrough on polio.

The whole series of press articles over Carver's successes were premature. Polio was so feared that many a parent confused any muscle problem as the onset of polio. Still, Carver's name was a powerful one. By the mid–1930s, people were coming to Carver for his peanut oil treatments. Carver moved cautiously, but every perceived success made the papers. Even medical schools wanted to test Carver's methodology. Carver loved the attention, and even though he could have, he never tried to profit from it. Letters came

Carver in his lab, 1929 (Alabama Department of Archives and History, Montgomery, Photograph Q4848).

requesting Carver's "special" peanut oil, but Carver lacked the mechanical presses needed to manufacture oil. People started to buy peanut oil where they could find it, often creating shortages. A number of old investors and new investors once again became interested in Carver's formulas. Carver remained on the edge of any commercial efforts but continued to be involved in treatments.

Carver's name in this work bought in physicians and others who saw commercial possibilities. Carver was more interested in getting some significant funding for his research. Publicity continued to mount as others looked for ways to profit and writers looked for stories to print. In 1937, a *Reader's Digest* article mentioning Carver brought the national spotlight to Tuskegee. A visit to Carver by President Franklin D. Roosevelt in 1938 started more rumors of Carver's "cure." The visit was little more than a photo opportunity for both men. In the end, Carver was unable to attract any major funding for his own research. The National Foundation for Infantile Paralysis showed no real interest in Carver's approach. Probably to appease the public, they funded a polio clinic at Tuskegee under the direction of a physician. All in all, this whole project was a mess of publicity, overoptimism, others trying to cash in, and failed hopes. Carver did promote the publicity and enjoyed the attention, but it was far from any success.

Austin Curtis acted as a gatekeeper as years passed. Curtis felt that many, for their own interests, often exploited Carver. Curtis also tried to improve press relations to better explain Carver's work. He often noted that Carver's faith was not unorthodox but very strong. Carver was not saying that God gave him recipes or products by putting them in the lab but attributed all in the universe to God. It was Carver's view that God was the source of all inspiration. This is why Carver did not oppose synthetics as unnatural. Synthetics were merely an extension of God's kingdom by his servant, man. While Carver believed the Bible to be the word of God and used it frequently to illustrate points in science, he was not a strict purist on interpretation of it. He was a spiritualist, believing the world was under the control of a higher power. Carver found appeal with all religions.

Austin Curtis, however, was not only a good chemist and assistant but also an entrepreneur. On arriving in Tuskegee, he was aware of the commercial possibilities of the Carver name alone. Curtis expanded on Carver's work with paints from recycled motor oil. Curtis once again worked on Carver's earlier formulas for hair tonic and massaging oils. With the financial aid of Carver's lawyer R. H. Powell and Carver's old YMCA student C. M. Haygood, a new company was formed to market "Peano-Oil" for hair and "Miracle Massaging Oil." The company was named Carvoline. Carver trusted all these supporters but declined direct involvement other than his name and recommendation. Some profits would go into a Tuskegee scholarship fund. The products were never fully successfully in Carver's lifetime, but Curtis sold the rights over the years.

Curtis went on to open his own business in Detroit, relying heavily on his work with Carver. Curtis expanded the product line with perfumes, work he had started as an assistant at Tuskegee. Curtis was an expert chemist in his own right and developed many chemurgic products. He had produced perfumes using flowers common in Alabama. Curtis had also represented Carver at the Ford Soybean Research Center and had developed a rug cleaner from soybeans. The products are still on the market today.

12

Chemurgy: A New "Political" Science

> "I believe the Creator has put oil and ores on this earth to give us a breathing spell.... As we exhaust them we must fall back on our farms which is God's true storehouse and can never be exhausted."
> — George Washington Carver

> "Farming is and always will be the foundation on which the economic growth of our nation depends."
> — Henry Ford

Henry Ford and George Washington Carver would meet for the first time at the 1937 Chemurgical Dearborn Conference. They knew well of each other, and Carver had been invited to but was unable to make the two previous conferences. Both men had been friends and admirers of Thomas Edison and Luther Burbank. Now in their seventies, they had seen the heights of national fame and suffered its negative side. They had been on equal paths of the science of chemurgy for decades. They had both led technological revolutions, but now the scientists and engineers that owed their careers to them saw them as over the hill. They were Victorians in a new age of science. To many, their day was over. While Carver had lived with discrimination, it was new to Henry Ford. However, a new movement was rising that would see them as legends to lead a new revolution. Carver and Ford would find a spring in their autumn with chemurgy.

The chemurgical movement would be the realization of their dreams. Both men represented the very core of the chemurgical movement—the elimination of waste, self-reliance, conservation, market-driven products, agricultural-based industry, industrialized farms, and American independence. Chemurgy was about a new political freedom found in American agriculture, and it would leave America free of foreign involvements. Ford was the promoter of soybeans, and Carver had his peanuts and sweet potatoes. They would now see the production of peanuts, sweet potatoes, and soybeans boom exponentially. Both men were agricultural engineers versus scientists. Their methodology was Victorian experimental based. They shared idols such as Thomas Edison and Luther Burbank. And these two men would converge to lead the chemurgical movement of the 1930s. Many of their ideas, which had been overlooked, would now become commercial successes. Their methodology would once again be meaningful in this new scientific frontier that lacked the theory of university scientists.

The chemurgical movement was every bit a political movement in its own right. It was a rebirth of the old Whig Party of Henry Clay and Abraham Lincoln. Chemurgy offered a new twist to the old Whigs by including agriculture as the political basis for American economic independence. The old Whig Party was a political party dedicated to the economic development of the United States through modernization, infrastructure, manufacturing self-sufficiency, and protectionism. Initially, the Whig Party was an economic movement free of any social issues. The Whigs stood for American economic might in every political approach. Its platform was designed by Kentucky congressman Henry Clay and was known as the "American System" in the 1820s. The Whig Party would have four presidents — William Harrison (1841), John Tyler (1841–1845), Zachary Taylor (1849–1850), and Millard Fillmore (1850–1853).

The Whig Party roots went back to Alexander Hamilton's "Treatise on Manufacturing" in 1794, believing in the need for manufacturing to be at the center of national economic policy. The Whigs enforced tariffs to protect the American worker and manufacturing companies but often at the expense of the farmers. Originally Jefferson and the Democratic Party of Jeffersonians opposed the Whigs of Hamilton because of the letter's emphasis on manufacturing. This farm-versus-industry divide would help bring down the Whig Party. It caused a political divide between farmers and manufacturers which the chemurgists could now close. For the chemurgists, America would assure its economic success through the union of farm and industry. It would close America's oldest divide between the Jeffersonians and the Hamiltonians.

While Carver is claimed as the first chemurgist and Henry Ford its first promoter, the word did not make its first appearance until 1934 in a book called *The Chemurgic Farm* by Dr. William Jay Hale. Hale made up the word "chemurgy" from the old Egyptian word "chemi," which is the root in "chemistry," and the Greek word "ergon" for "work." Thus "chemurgy" was "chemistry at work." Hale noted that the pronunciation is "Kemurgy," with a soft accent on the first syllable (like metallurgy). For Ford it was the perfect word right out of the syllable-based approach of his *McGuffey Readers*. And it was Ford who made the word appear in national print with his chemurgical Dearborn conference in 1935. Chemurgy would become the term for the commercial application of farm products in industry. The 1935 conference would bring together a new movement.

Dr. William Hale's journey to Ford's first chemurgical conference was a fascinating one that would bring together two of the other major promoters of chemurgy at Dearborn in 1935. William Hale was born in 1876 in the farm community of Ada, Ohio, a son of a minister. Like Ford and Carver, he grew up with the *McGuffey Readers*, schoolboy experiments, and a love of nature. He went on to Miami University in Ohio and later graduate school at Harvard to study chemistry. He became a professor of chemistry at the University of Michigan. At Michigan he met and married Helen Dow, the daughter of the founder of Dow Chemical. Hale became the chief chemist at Dow and an industrial leader in chemistry and chemurgy.

Hale was recruited by Francis Garvan of the Washington-based Chemical Foundation to speak before Congress in defense of tariffs for the chemical industry in 1921. It was the same committee that had heard George Washington Carver. It might well have been that the beginning of chemurgical movement was born in that committee room.

Like Carver with peanuts, Hale got protection for American chemicals in the Fordney-McCumber Tariff Act. Just as important was that organic chemicals and related industries gained a high degree of protection to operate in the United States. The congressional testimony for that tariff act would bring together a third player in the early chemurgical movement, Dr. Charles Holmes Herty, the president of the American Chemical Society. Herty was also working on the Chemical Foundation that had been set up after World War I as a clearinghouse for seized German patents. The threesome of Garvan, Hale, and Herty would take a camping trip near Washington and become lifelong friends, meeting every year in Midland, Michigan (home of Dow Chemical), to hunt, camp, and discuss chemurgy. These men saw Carver and Ford as the iconic leaders in this new movement, which was a fusion of science, industry, agriculture, and politics.

At the time, William Hale was the Carver of the North, seeing potential chemicals and products from farming. Hale believed in the industrial farm vision of Ford and Carver, although he was more industrial chemist than chemurgist. Hale saw the chemurgical movement as scientific at its root, but acknowledged the economics. The timing also was perfect as surplus crops continued to bring crop prices down in the 1930s. However, the chemical industry, including Dow Chemical of Hale's father-in-law, was hesitant to invest equipment in agriculture that was dominated by the commodity market and world governments. Hale was having problems even getting his ideas published in magazines and journals. It was Henry Ford that finally got Hale published in his *Dearborn Independent* in 1926 under the title, "Farming Must Become a Chemical Industry." Then in October of 1926, Hale made a national radio broadcast on the *Ford Sunday Evening Hour*. Hale's ideas would soon become part of the chemurgical movement. It brought farmers, chemists, and industrialists into a unique alliance. It would also bring together the key players in the movement who had been working individually such as George Washington Carver, Homer Herty, and Wheeler McMillen.

Wheeler McMillen, associate of editor of *Farm and Fireside*, had been traveling across the country interviewing scientists on this very topic. One of those experts had been George Washington Carver. Through his editorials, McMillen was able to change minds to bring the farm-industry alliance together. McMillen was a true advocate of the farmer and the chemurgical movement. He was less interested in the industrial uses per sec as long as farms were at full capacity. McMillen was of the pure Jeffersonian mode that agriculture was the real wealth of the nation. Like Ford and Carver, McMillen saw farming as the social fabric of the nation. He discussed the alternative uses of plants that Hale, Ford, and Carver had been talking about for years. He also started to push farm-based distilled alcohol mixed with gasoline as fuel.

McMillen brought the national press into the movement. This was of great interest to Henry Ford, Thomas Edison, and then Secretary of Commerce Herbert Hoover, whom McMillen interviewed as well. Ford invited McMillen to visit his many efforts to combine industry and agriculture in Dearborn, Michigan. It also caught the imagination of Francis Garvan who put the printing press of the Chemical Foundation behind the idea. Garvan won over Irene DuPont, vice chair of E. I. DuPont Nemours Company, and the board of

12. Chemurgy: A New "Political" Science

directors at Dow Chemical. McMillen is often credited with congealing the chemurgical movement.

The full movement reached a peak at the 1932 World's Fair with Ford's famous soybean exhibit. Hale also published a best seller, *Chemistry Triumphant*, for the World's Fair. The book proposed farm chemicals as a solution to the nation's overproduction. Still, just as the momentum was building, the political winds shifted with the election of Franklin D. Roosevelt. Roosevelt's advisors, seeing the economic depression coupled with a farm surplus and low prices, were formulating a plan to destroy surpluses and pay farmers not to plant. The chemurgists looked, as Carver had with peanuts, to find new industrial uses for crops to increase demand. Corn would reach a low of ten cents a bushel in early 1932. This bottoming out of corn prices created a political farm movement known as the "power alcohol."[1] It resurrected the Ford ideal car-fuel mix of 40 percent alcohol and 60 percent gasoline.

Henry Ford had suggested the use of grain-distilled ethanol in gasoline from the first days of the Model T. However, the opening of the oil-rich wells in Oklahoma flooded the market with cheap gasoline prior to World War I, ending this early push of Ford's. After World War I, there was a glut of low-priced alcohol, and a price increase in gasoline by the prohibition movement had idled the distilling capacity. Standard Oil entered the alcohol mixed gasoline business briefly in 1922. Using excess alcohol, Standard tested a 25 percent mix. But once again, new oil discoveries in California and Texas, and improved refining technology, put cheap gasoline on the market, ending the alcohol mix. While alcohol/gasoline mixes offered the elimination of engine knock, the oil companies found that adding tetraethyl lead to be a cheaper alternative. Lead tetraethyl would be a major toxin for decades until banned in the 1970s. William Hale had teamed up with Ford and Leo Christensen of Iowa State in the 1920s to push the power alcohol movement, but it was difficult to match the cost of five cents a gallon for gasoline at the time.

The fall of corn prices in the earlier 1930s rejuvenated the movement. Politicians in the Corn Belt seized on it as a solution to depressed farmers. Corn had been the favorite grain for power alcohol, although Carver would argue that the South could make more from sweet potatoes, white potatoes, Jerusalem artichokes, and even waste wood chips. Carver's old school, Iowa State, would take leadership in chemurgical research to grow our fuels. Congress and corn states such as Iowa pushed legislation that called for 10 percent alcohol in the gasoline. The big oil companies organized the opposition to defeat these efforts by arguing that engines would be hurt by it. Ford put his support with the Corn Belt, but he was alone in industry for alcohol power fuels. Chemurgist Francis Garvan of the Chemical Foundation added his support with publicity. Chemurgist William Hale added his support with Ford and Garvan in his book, *Farm Chemurgic*. Chemurgists found some success in winning over other auto manufacturers. Auto executives such as Charles F. Kettering, vice president of research for General Motors, joined the power alcohol and chemurgical movement. The power alcohol movement was engulfed in the chemurgical movement as Henry Ford, Francis Garvan, and William Hale planned the nation's first chemurgical conference in 1935. The farm and industrial depression of the

1930s was the perfect environment in which to launch the movement. The farmer and industrialist stood together in this economic depression.

Henry Ford would bring the movement together with his 1935 chemurgical conference in Dearborn, Michigan, in 1935. Ford would bring together several streams of chemurgy in 1935. First would be those close to Ford such as rubber latex, soybean usage, and power alcohol. There was also a stream of idealistic chemurgists, such as Wheeler McMillen, with an ideology of conservation-ecological chemurgy, which is analogous to the green movement of today. There were the farm proponents wanting more farm product usage and better prices. There was a large group of chemists and scientists looking to expand into this new area which had brought William Hale to the movement. There was even a health stream of chemurgists wanting to replace meat and processed foods, and a stream of environmentalists. The economic stream looked to chemurgy and Francis Garvan as a solution to the Great Depression by utilizing America's farms and industries. There were small but radical segments such as those wanting to legalize hemp and lower taxes on alcohol. Finally, there was a political stream opposed to the New Deal practice of paying farmers not to produce. Ford's genius was to bring these streams together under the banner of American independence and economic freedom.

In 1935, Ford Motor was the corporate leader in chemurgical products, and Ford used the conference to highlight that leadership. In that same year, Ford Motor paid over $27 million for agricultural products to manufacture over a million cars. Ford estimated that the total acres needed to grow these products were 1.2 million, or the equivalent area as the state of Rhode Island.[2] Ford was utilizing cotton, wool, wood, flax, tung oil, jute, beeswax, pine pitch, castor oil, hemp, flax, sisal, corn, and soybeans. He had twenty parts in soybean plastics such as the distributor cap, accelerator pedal, and steering wheel. Other uses of soybean products included paints, plywood glues, pressed board, and ink. His soybean "wool" was being used in upholstery. In 1936, Ford added 400 plastic molding machines to the Rouge plant to make soybean plastic parts, including a bumper for his new Ford V-8. Ford Motor got 90 percent of the needed soybeans from a 200-mile radius of Detroit. Ford Motor in 1935 was experimenting with hubcaps and other trim made out of soybean plastics. The company was also producing stearic acid from soybeans to make experimental synthetic rubber.

Ford's 1935 conference would birth the alliance of industries, farms, chemical organizations, trade associations, and universities into an organization known as the Farm Chemurgic Council. The 1935 Dearborn Conference was to select a number of committees to give the chemurgic movement the infrastructure it lacked. Francis Garvan of the Chemical Foundation was named chairman. The official sponsors were the American Farm Bureau Federation, the National Grange, and the National Agricultural Conference. The conference had major executives from Ford, General Motors, Dow Chemical, E. I. DuPont Nemours, the American Chemical Society, and the United States Chamber of Commerce. Major universities such as the University of Chicago, the University of Michigan, and Harvard were represented. There was a small group of politicians, but both national parties were following the proceedings closely. The mere fact of these powerful executives coming together on anything was troubling to politicians.

With the mixture of Irish and highly opinionated men such as Ford, Francis Garvan, and William Hale, it was not surprising the conference took on some perceived enemies. These included big oil, banks, foreign countries, and New Deal Democrats. Garvan was a strict nationalist and believed in economic isolation and self-sufficiency, in direct opposition of the free-trade policies of the Roosevelt administration. The chemurgical movement did escape being captured by the politicians, forcing some concessions from both sides. New Deal Democrats held to their policies of paying farmers not to produce and destroying surplus but did pass legislation for funding more agricultural research in chemurgical projects. Wheeler McMillen played the role of mediator with the Roosevelt administration trying to make smaller deals when possible. It even brought in some surprising politics. Southern Democrats often broke ranks with New Dealers to expand chemurgic efforts. Democratic senator Theodore G. Bilbo, one of the nation's biggest racists, did combine with the support of chemurgists such as Ford and Carver to pass legislation to expand Federal Agricultural laboratories and experimental stations to even black areas. Bilbo represented the cotton-dependent state of Mississippi. Power alcohol from agriculture was a main plank of the chemurgists, but one opposed by New Dealers and the oil companies. By the end of the 1930s, chemurgists lost the battle for power alcohol with the New Deal policies to limit and destroy surpluses.

The Depression and the New Deal policies affected all farm products. Even the alternative southern crops such as the peanut suffered from the economic decline. In 1932, peanuts dropped to under three cents a pound. The first response of Congress was to offer low-cost loans to peanut growers, but this did little for the small farmers and not much for the large farmers. The New Deal Congress of 1933 tried to put in direct price supports for peanuts, but this was a total failure. The New Dealers saw the farm and peanut problem as a supply/price issue. So the next step was to restrict production and pay farmers not to grow. To finance the payments, taxes were placed on the peanut processors. The program ended as the Supreme Court ruled the processing tax unconstitutional. The government then moved to pay farmers to plant other crops, but again with little success. The chemurgists, of course, argued that the real issue was a demand problem which required research into new applications. As a type of compromise, the government tried to force more acreage from eating peanuts to oil. Peanut oil represented only a small percentage of the acreage, but again vegetable oils in general were depressed markets. It took World War II to finally increase demand; but looking back, chemurgists would argue that government market control did little.[3] In fairness, the chemurgists appear correct in their solution to expand demand, but it was a longer-term solution (other than ethanol production) that would have offered limited fast relief which many demanded. Chemurgy, however, did make a contribution to improving farm income during the Depression.

The other part of the story of chemurgy during the 1930s was its steady success in increasing markets for farm products. By the end of the 1930s, more moderate leadership of the chemurgical movement took over more in the vein of George Washington Carver. In 1938, Dr. H. E. Barnard took over the movement with the deaths of Francis Garvan and Charles Herty. Barnard had been chairman of the 1937 conference that brought Carver and Ford together. He stated the chemurgical mission in 1938:

Carver and Ford's first meeting in 1937 at Dearborn Inn (Alabama Department of Archives and History, Montgomery, Photograph Q4846).

We must subject every promise in the field of chemurgy not only to the scientist but to the economist and in many cases to the sociologist. For it is not enough to find out how to make something out of wheat flour; it is necessary to know the effect of the diversion of wheat to the new product might have in the broad field of economics. And it is equally necessary to know what the diversion of the crop to new uses might mean in the way of new social trends. Chemurgic farming then must travel a narrow road, avoiding excursions on the one hand which are economically unsound and on the other the creation of beliefs that promise of large industrial markets for surplus crops will soon be realized.[4]

Of the four major chemurgical Dearborn conferences (1935, 1936, 1937, and 1938), the dominant session of all was on power alcohol. The first major victory of the chemurgical movement was the power-alcohol plant at Atchison, Kansas, in 1936. The Atchison plant had been advertised heavily by all the outlets of the Farm Chemurgic Council. The ethyl alcohol was made from corn to blend in percents of 5, 10, and 15 with gasoline (typical of Midwest gasoline blends today). It would be called "Agrol." Gasoline was selling for seven and a half to eight and a half cents a gallon. The Atchison alcohol was produced at twenty-five cents a gallon. Blends would be sold at nine cents and above as a premium high-octane antiknock product. The Farm Chemurgic Council believed that as volume increased, the price of power alcohol would decrease; but they never achieved a much lower cost than twenty-three cents a gallon. Originally, the Farm Chemurgic Council believed fifteen cents a gallon was possible. The plant ran mostly from corn, but in an effort to reduce costs, sweet potatoes, white potatoes, and Jerusalem artichokes were tried. At its zenith, Agrol was sold at over 2,000 gas stations in the Midwest.[5] At the same time, Ford was having his labs look at hemp as a base for alcohol production. He reasoned that hemp produces four times the cellulose per acre of trees and grows faster. Like scientists today, Ford believed that alcohol from cellulose was the long-term solution in bringing down cost.

Chemurgists such as William Hale saw alcohol as the king of chemicals and the potential root of all industries. Alcohol could run our cars and be the raw material for plastics and synthetic rubber. Chemurgists, however, were pushing for an emergence of a Midwest alcohol industry to be driven by small distilleries. One problem was the East Coast sugar refineries, which were the main source of industrial alcohol. The sugar industry had a powerful lobby. These old industrial networks worked against the chemurgical dream of agriculture and industry. As politics closed off power alcohol for cars, chemurgists looked to fiber production. Ford was already leading the way with synthetic wool from soybeans.

Chemurgists were chemists and were not threatened by the emerging "synthetic" fibers of the 1930s. They looked to work with synthetics versus looking for political acts to restrict them. Carver had once said, "The great Creator gave us three kingdoms, the animal, the vegetable, and the mineral. Now he has added a fourth — the kingdom of synthetics."[6] Chemurgists were demand oriented, refusing any artificial restrictions of the domestic production of chemicals or man's creativity. Chemurgists looked at farm products as falling into three broad material categories — carbohydrates (sugar and starch), cellulose (cotton and wood), and oils. Chemurgists did not see plastics or synthetics as a threat nor as a complementary industry. This was consistent with Carver's view, as synthetics were a natural part of God's world. Rayon was at its root regenerated cellulose and could be made from wood and even cotton (both cellulose). Rayon and cellophane are cellulose, as is cotton, so chemurgists saw them as one version of the same. Cellulose acetate had been a growing plastic from 1932, and had been made into rods, tubes, sheets, and even photographic film. It was also used in automobile safety glass. Prior to his death, Ford was ready to commercially release a synthetic soybean wool called Azlon. Ford engineers were even marketing a sports shirt made of 50 percent rayon and 50 percent Azlon, which would have the warmth of wool and the softness of cotton.

Cellulose plastics were also being made by nitrating cellulose from cotton, wood, hemp, and other plants. These plastics known as celluloid, which had become popular in billiard balls in the 1870s, were replacing ivory. Celluloid had been popular as a bone and ivory replacement. It had first been used as a movie picture film, but it was highly flammable and was replaced by cellulose acetate in the 1930s. Always chemists at heart, chemurgists were studying the new and maybe first real plastic, known as Bakelite. Bakelite was the product of the home laboratory of Leo Baekeland in 1907.

Not surprisingly, the 1936 Chemurgic Conference in Dearborn gave their "Pioneer Cup" to Leo Baekeland as the "father of the plastic industry." Baekeland had been inspired in 1863 by a $10,000 reward for anyone who could find a substitute for ivory billiard balls. Bakelite was the product of the reaction between phenol and formaldehyde. Phenol was a coal-based chemical and a by-product of coke making. Formaldehyde was truly a chemurgical product made from wood alcohol (methanol). Bakelite was a hard thermosetting plastic, which was flameproof. By 1915, Bakelite could be found in telephones, electrical insulators, knife handles, washing machine agitators, radios, and combs. There was even concern of a Bakelite shortage because of the success of Thomas Edison's phonograph, which used Bakelite records. Edison had started a research project to find a substitute. Bakelite truly launched the plastics age, but chemurgists looked to join it, not oppose it. In 1919, there was a huge surplus of phenol from the production of explosives, which had little application other than aspirin and Bakelite. The price of Bakelite dropped as its use as an insulator boomed in the automotive and electronic industries. The fear became that once the surplus was gone, phenol and Bakelite prices would skyrocket. Chemurgists began the work for a cheaper process to manufacture phenol. Another chemurgist, William Hale, would invent the Hale-Britton process for phenol production which also reduced the price of phenol.

While chemurgists worked with the plastics movement, coal/oil-based plastics would bring down chemurgical research with cheap prices from oil and coal at the time. Coal tar and phenol from coke plants were early resins and plastic material. Ford and many chemurgists saw a harmony between plant and fossilized plant sources. Ford was in the coal and coke-making business and needed to sell by-products such as phenol to be made into plastics. Ford Motor took leadership in the use of coke-making by-products over the great Pittsburgh steelmakers. Coke making offered great savings in what was once called waste products. Still, chemurgists preferred renewable materials to coal and oil-based ones. Ford Motor, of course, through its wood-processing plants, was a major producer of wood alcohol and thus formaldehyde for plastic production. Ford, however, was more focused on plastic that came from renewable plants.

Ford was pushing the use of soybean-based reinforced plastics and was manufacturing car parts at his River Rouge plant. The Fairchild and Hughes companies were experimenting with soybean plastic aircraft in the late 1930s as well. Ford was also following Carver's lead with the manufacture of shredded wood impregnated with vegetable resins to make compressed wood and recycled wood products. By 1942, the plywood industry was using 60 million pounds of soybean glue, representing 85 percent of the plywood glue used. Carver had produced wood glue from sweet potatoes as a binder for pressed wood and as a binder for other composite materials. Both Ford and Carver had developed high-strength composites of plant resins and natural fibers.

Hemp with soybean resin produced one of the strongest materials for car production. Hemp with resin had been used to strengthen the famous leather Swedish cannons of the 1600s. Hemp with gutta-percha had been used in the late 1850s to strengthen and protect the transatlantic cable. Carver had years of experience with composite materials such as cotton-reinforced asphalt and fiberboard. The development of reinforced plastics would be a project of both Ford and Carver's in the late 1930s. The two men worked a number of veggie-plastic composites from cotton, corn stalks, flax, and wood pulp. For Carver and Ford, legumes offered the greatest potential plant-based resins in reinforced materials. Peanuts and soybeans offered not only industrial materials, but also a replacement of animal fats, oils, and other products. Chemurgists set the foundation for the future of reinforced materials in industry.

Ford used hemp and resin to produce his famous farm-growth car. Hemp fiber offered strength, and when mixed in soybean resin or phenol resin (from his coke plant), it produced a very light, strong panel that could be molded to size. Ford looked to take an ax to his soybean car to prove a point. His hemp-soybean panels could take a blow ten times as great as steel without denting. These hemp-soybean composites took 1,000 pounds out of the weight of the car (about a third of the total weight), dramatically improving car mileage. The aircraft industry took notice in the late 1930s and started experimenting with hemp composites and even wood composites. Early in World War II, the aircraft industry used glass fiber with phenol resin to create "fiberglass." For Ford's farm-growth car, he also used hemp for the upholstery, hemp oil for lubrication, hemp oil in paints, and he even used hemp alcohol to power his car. Ford and Carver even suggested that hemp-reinforced concrete could help America's highways.

While Ford and Carver were working to eliminate cows and dairy milk, other chemurgists were working to create fibers and plastic from cow milk. Casein was a milk protein that could be mixed with formaldehyde to form a plastic material. It was the basis of a product known as Galalith (Greek: *gala* = "milk," *lithos* = "stone") to make buttons, combs, lampshades, and pen cases. In the 1920s it was becoming a popular replacement

Ford's 1942 soybean car (from the Collections of The Henry Ford, Benson Ford Research Center, Photograph THF95637).

for bone and ivory. Surgeons had tried limited amounts for bone replacements. Even George Washington Carver had used casein-based paints for artwork. Casein paints remained popular until they were replaced in the 1960s by acrylic paints. Casein plastics were helping to reduce the popular slaughter of elephants for ivory.

Milk casein was a hot topic at these chemurgical conferences because of the milk surplus of 1937–1938. There was one cow for every five Americans, producing 100 gallons of milk for every living American. There was an excess of butter being stored by the government at the taxpayer's expense. The government was looking to put in controls while chemurgists were looking at more uses, such as a wool substitute out of Europe, using milk. While the government looked to reduce or even destroy supply, chemurgists wanted to utilize the surplus. The wool substitute was a casein-based product known as "Lanital." German and Italian chemists had produced various types of this chemical wool. They had mixed Lanital with wool to make a commercial product.

Chemurgists wanted to work with farmers and the utilization of any product surplus versus artificial restrictions. In the 1930s, they weighed fully into the politics of margarine versus butter. Margarine had been part of a long war with American dairy farmers. The battle of margarine versus butter was one of the oldest in American farm politics and one that Carver was involved in for decades. It started in Europe in the 1860s with the great Victorian motivator of technology, Napoleon III, who offered a large prize for anyone who could discover a substitute for butter. Napoleon wanted to make a butter substitute available to his military and poorer subjects. The product was based on an earlier French chemist's work with identifying margaric acid in animal fats. This led Mege-Mouries to formulate margarine using suet, minced sheep stomach, chopped cow udder, and a little milk. This concoction took the prize but lacked popularity in America. However, in the 1880s, Americans improved the taste, and margarine stored better and spread better. The dairy lobby in Congress passed the Margarine Act of 1886, which put on a two-cent tax and licensing fees. Still, it didn't stop homemade options such as those developed by George Washington Carver in the early 1900s. Animal fat stored better in the South and had been used in cooking since colonial times.

In 1905, the process of hydrogenation was developed, which allowed vegetable oils to be hardened into a butter-type consistency. Crisco, a soft hydrogenated vegetable oil product, hit the market in 1911 as a cooking substitute for butter and lard. Crisco used cottonseed oil. Harder oil products started to hit the market while Carver was experimenting with homemade products. He was developing margarine-type products using peanut and soybean oils. These hydrogenated oils stored much better in the hot South. The dairy lobby once again tried to stop the growth of butter substitutes with laws that restricted coloring of margarine with yellow dyes. Minnesota even passed a law to have manufacturers color it pink. Of course, Carver had been making margarine from peanut and soybean oils for years. His early formulas were lard based, but he moved to hydrogenation, expanding his research. Chemurgists got involved, showing farmers that vegetable oils could be a plus for them in the 1930s. America needed to increase production of oil crops. By the nature of the problem, chemurgists were dragged into the New Deal farm policies of reducing production and planting to keep prices high.

The issue was that most vegetable oil was being imported and production techniques had to be improved. Even cottonseed oil was mostly imported. At the 1935 Dearborn conference, chemurgists reported that America used 3,000 tons of imported vegetable and animal oils every day. Coconut, palm, cottonseed, soy, olive, peanut, tung, and corn oils were all being mostly imported because of their low price. The "tropical" oils of coconut, palm, and tung had almost no domestic sources. The soap and paint industries were two of the major users of these imported oils such as palm and coconut. Carver had shown that peanut made an excellent substitute for coconut oil in soap making, but cost was the problem. In 1935, over 750 million pounds of tropical oils and 200 million pounds of cottonseed oil were being imported. Tung oil offered the best hope for growing tropical oil in America. In a 1920 experiment, Ford was growing gung trees in the South to produce a domestic source for it. Imported tung oil dominated the market for linoleum, printer's ink, and certain paints. The better hope of the chemurgists was to expand domestic-sourced oils such as soybean, flax, peanut, cottonseed, and corn.

Carver and other chemurgists had increased the number of products that used vegetable oil, such as linoleum and paint (20 million pounds per year). Both Carver and Ford had led the push for more uses of peanut and soybean oil, but now more domestic crop production was needed badly. America was importing 16 million pounds of it in 1937. Chemurgists estimated that America needed to double its peanut acreage and triple its soybean and corn acreage. In addition, chemurgists were experimenting with sesame, flax, and sunflower oil. One problem continued to be the economics of dumping tropical oils at low prices. Chemurgists pushed for government tariffs using the old Whig/Republican logic of protecting infant American "industries" such as vegetable oil production. Chemurgists also looked to Carver to find new uses for specific oils and to Ford to expand planting and reduce production costs.

Chemurgists of the period were even working on synthetic rubber made from alcohol, although synthetic rubber products from oil were going to market in things like shower curtains, and the market was growing in the late 1930s. The rise of Germany in the 1930s again brought fears of our strategic rubber supplies from Asia and South America being cut off. In 1935, America consumed 450,000 tons of plantation rubber. While Ford, Edison, Carver, and others had looked for a new plant source for rubber, chemists had unlocked the chemistry of rubber in the 1920s. Synthetic rubbers made from ethylene had hit the market in the 1920s under names like koroseal, Thiokol, and deprene. Most of these came from the work of a Catholic priest and chemist at Notre Dame. DuPont Chemical had improved on it and introduced it as neoprene rubber made from ethylene, which is a gas from the oil-refining process. Still, by the end of the 1930s, there were less than 5,000 tons a year of synthetic rubber being manufactured.

Cheap oil and the oil companies tried to keep the market with their synthetic rubber known as butyl rubber, which was made from two refining gases, isobutylene and isoprene. Buna-S synthetic rubber was made from butadiene and styrene, which could be made from oil, coal tar, or alcohol. Russia was making it from alcohol, and Germany was using alcohol and coal. Standard Oil had made an agreement with German manufacturers not to expand its synthetic rubber production if Germany agreed not to compete in the

American petroleum market. The Germans were not only making rubber from coal but also gasoline. Chemurgists were fighting for corn-alcohol-based Buna-S rubber against the oil cartel.

As the 1940s and war approached, there was a battle among politics, farm interests, big oil, and the chemurgists for patents and legal rights. The chemurgic lobby got a major bill passed for the development of the alcohol-based rubber industry, but it was vetoed by President Roosevelt. Finally, the government stepped in to take control because rubber was a strategic war material. The government proclaimed that 800,000 tons of synthetic rubber a year were needed. It declared that of the 800,000 tons, 60,000 tons were to be Standard Oil's butyl rubber, 40,000 tons were to be DuPont's neoprene rubber, and 700,000 tons would be Buna-S rubber. However, only 15 percent of the Buna-S rubber was to be made from alcohol. President Roosevelt assigned a "rubber czar" to monitor the production. It was a real blow to the chemurgists pushing for corn-, wheat-, and potato-based alcohol to be used.

The real irony would be that when the war came, it would be alcohol-based Buna-S rubber that saved the nation. The petroleum industry had trouble bringing the rubber production online to meet the government quotas. In fact, petroleum-based rubber was in short supply during the early years of the war. Alcohol-based processes came onstream faster, and the chemurgical company of Reichhold Chemicals improved the process's efficiency. Reichhold called their rubber "chemurgic rubber" or "Agripol." By the end of 1943, 130,000 tons of alcohol synthetic rubber were produced for *77 percent* of the nation's total; and in 1944, alcohol-based rubber reached 362,000 tons.[7] Without alcohol rubber, America's war effort would have been crippled. Every Sherman tank required 1,000 pounds of rubber, all of the nation's 30,000 bombers required 2,000 pounds each, and every battleship needed 160,000 pounds. The army required 45 million pairs of rubber boots, and the air force needed 1.4 million rubber tires for its fighters. Lost to history is this critical contribution of chemurgy to America's winning the war.

Nylon was a true synthetic from coal and natural gas; but here again, chemurgists looked at it as a complementary material to be mixed with rubber and other cellulose products. E. Barnard even pronounced DuPont's nylon as being chemurgic, made of fossilized plants, which had been the basis of Carver's early classification. Carver and Ford had been looking at fibers for years, and chemurgists had taken up the research in the 1940s. Synthetics like nylon had their place, but there was a growing interest in hemp, flax, and grass fibers such as ramie. These natural fibers had amazing tensile strength. Ramie was a Chinese grass that could be grown in the South, and Carver had experimented with it. Ford also had experimented with it at Ways, Georgia, and had developed car parts from it. Ramie was little known, but it was used in the wrappings of mummies. Ramie was an amazing fiber with eight times the strength of cotton, four times that of flax, and three times that of hemp. Ramie cloth washed easily and dyed better than cotton. It had the feel and look of silk and was great in hot climates.

Ramie flourished on poor soils, and yields equaled cotton, a fact that had originally interested Carver in it. Ford had started to add ramie fiber to strengthen his plastic car panels. He increased his fields of ramie at his Georgia experimental farm. Ramie yielded three crops a year. What was holding up ramie was the difficulty in harvesting and processing it. Efforts to produce or "gin" its staples had started in the 1890s with little success. Chemurgists had been working hard in the 1940s to find the mechanical methods needed to make the fiber useable. They never fully found a cost-effective method.

They also tried bamboo from the grass family with a little more success. It is only recently that "bamboo rayon" is gaining markets. "Bamboo rayon" is dissolved bamboo, then drawn like rayon into a fiber. Bamboo had also been studied by Carver and the chemurgists for use in the making of paper. Bamboo could produce as much as six times the pulp of southern pine per acre. Another fiber that Carver had studied early on that became popular in the 1930s chemurgical movement was yucca. A strong fiber could be made from yucca, which was a wild cactus and was as strong as hemp. During shortages of jute in World War I, yucca fiber was used to make rope as Carver had done years before.

13

The Meeting of Ford and Carver

Carver, who would be called the nation's first chemurgist, didn't join the movement formally until 1937 at a Dearborn Conference sponsored by Henry Ford. Carver had been invited to earlier conferences in Dearborn, but health problems had prevented him from attending. Earlier in 1937, George Washington Carver attended the Farm Chemurgic Conference in Jackson, Mississippi, from April 12 to 14, 1937. This conference had been sponsored by the State of Mississippi, and it was a real breakthrough to ask a black scientist to speak. Carver joined the governor of Mississippi and distinguished chemurgists such as William Hale of Dow Chemical, a founder of the movement. In Carver's speech, he dealt with the relationship of the chemicals God had given to the new synthetics that he had guided man to develop. Carver noted that God had really given us four kingdoms — animal, plant, mineral, and synthetic. Like most chemurgists, Carver saw a harmony in organic and inorganic chemistry as well as the new synthetics.

Carver and Ford had been on the same chemurgic path for decades. They were destined to meet. They were like soul mates in so many ways. Both were a bit shy and soft-spoken in manner. They were frugal, thrifty, and conservative in personal habits, which they translated to their overall careers. They shared many idols such as Thomas Edison and Luther Burbank, were both naturalists and conservationists, and they both had a love of nature. While they came from different careers, they both believed in an alliance of industry and agriculture for economic advancement. Both men had made major contributions to the advancement of blacks in the nation. It seemed fitting that Carver's speech at the 1937 Dearborn conference where the two men would meet was titled "What Chemistry Means to My People." They had much to talk about in soybean, hemp, flax, peanut, sweet potato, and plastic research. They shared the same approach to the education of American youth. Their quiet demeanor could hide their goal-oriented and hard-driving nature.

Ford had been reading the bulletins of George Washington Carver and had started contributions to Tuskegee back in the early 1900s. They had started an exchange of letters in 1934, and Ford had invited Carver to his first chemurgical conference in 1935, which Carver had been unable to attend. Letters between the two and their agents continued. Carver's health had started to decline after 1934, but Ford wanted Carver at his 1937 conference. Carver had become symbolic of the chemurgical movement. Ford was particularly interested in the potential of sweet potatoes for glue and binding wood products. After reading Carver's bulletins, Ford had directed more research in Dearborn and his experimental station in Georgia. Ford wanted to team up with Carver and bring him to Dearborn. He hoped to get him to stay and work in the Ford labs on chemurgic projects, in particular, soybeans.

Henry Ford would ensure that a special car was used to transport Carver to this 1937 Chemurgical Conference. He would arrive at the special Greenfield Village station and be taken by car a short distance to Henry Ford's exclusive Dearborn Inn. The Dearborn Inn had been built by Ford to accommodate such guests. Ford would be Carver's constant companion for the three days. The press was fascinated by these two elderly (both in their seventies) geniuses meeting. Ford took Carver to his industrial museum and on to his village industries. In the museum, Henry Ford had the best collection of farm machinery as well as spinning and fiber-processing equipment to show Carver. The conference itself, with its days of talks, would bring many attendees today. Topics such as ethanol cars, alcohol/ethanol fuels, hemp-reinforced plastics, biofuels, green plastics, recycling, and many themes of the green movement today were discussed. The press crowded the halls of the conference.

Ford had only a few months earlier attended the funeral of his vagabond friend Harvey Firestone. He now was without his two creative muses, Edison and Firestone. Ford hoped that maybe Carver could fill that role. Most important, Ford took Carver to his "soybean" research center in Dearborn. In Ford's historical Greenfield Village, he had twenty-five acres of various experimental varieties of soybean. Ford had a chemical plant at Greenfield that was producing over ten tons of soybean oil daily. Both men had done extensive research on soybean products and shared a soybean ice cream cone at the plant. They would spend a few hours here, like kids with a Christmas chemistry set. The press enjoyed the meeting as much as Ford did. Carver was to be a headliner at the conference dinner.

After Carver's death in 1943, the editor of the *Detroit Free Press* would recall Carver waiting outside the famous 1937 dinner waiting to speak: "When I saw him first that day he was sitting out in the corridor by the door of the banquet hall. The place was filled with leaders of industry and science. He was to be the speaker of the evening. But he did not enter. He had dined alone in his room. I asked him why he was not at the speakers' table. 'It's nicer out here,' he smiled, 'some people just do not understand. They'll call me when they are ready for me.' And so we sat and chatted and I forgot all about the dinner and other speakers."[1] As always, Carver won over the crowd. He was an encyclopedia of chemurgical applications. Carver brought chemurgy down to a personal level that appealed to the farm segment in the conference.

The conference would highlight the development of an experimental hemp station in Minnesota as well as Ford's own work with hemp. Another point of an agreement for Carver and Ford would be that of industrial hemp as a crop of the future. Hemp (the industrial variety of marijuana) was the darling of chemurgical materials prior to sweeping legislation in 1937 against growing any varieties of *Cannabis sativa*. Just prior to the chemurgists losing this battle with the government, over 110 million pounds of hemp oil were being used for paint and varnish. Both Henry Ford and George Washington Carver supported the industrial use of hemp, with good reason. It was not only stronger than cotton, but it required far less fertilizer. Harry Hans Straus was the speaker at this conference who would report on the success of hemp manufacture in Minnesota before the government shut down the operation. Ford engineers had also done work in Michigan on using hemp oil as a replacement for linseed oil since they both had similar drying properties. Ford and Rudolph Diesel saw hemp oil as a potential fuel as well.

By the mid–1930s, Ford was using hemp as the fiber in his soybean plastic doors and car parts. Hemp had once been an important industrial fiber in America. During the Revolution and early 1800s, hemp was used in sails, ropes, wagon canvas, and industrial bags. Carver had suggested its use in fiberboard because of its strength. Before the Civil War, it had been used in Levi's famous jeans because of its wear resistance. Hemp was strong, tough, and moisture resistant; and at 85 percent, it has the highest cellulose content as a fiber. Before the Civil War, hemp made up 90 percent of world paper production until the cheaper wood pulp process replaced it. Ford had produced fine paper, as in the past, from his hemp crops. Carver loved industrial hemp because all parts of the hemp plant offered industrial promise. For decades it was the major crop of Kentucky and was protected by tariffs. Henry Clay, the father of home-centered agriculture and manufacture and the founder of the Whig Party of American economic development, had introduced hemp tariffs in the 1820s as part of his system of protective tariffs. Hemp production was extremely labor intensive, and the end of slavery increased its cost. Also, cotton became cheaper with the invention of the cotton gin. Jute and abaca became low-cost substitutes by the twentieth century for hemp's use in naval sails and rope, and nylon became a synthetic competitor.

Like Ford, Carver had been an early supporter of a return to hemp production. Carver wrote, "It is grown in Kentucky and many sections of the U.S. for its strong fiber, which makes an excellent quality of linen cloth, thread cordage, etc…. The seeds are also sold in large quantities for making bird-seed mixtures."[2] Hemp had proven the major base for good paper for centuries. Hemp paper lasted hundreds of years longer than that made from wood pulp. In 1937, hemp paper was still produced for the world's finest paper books such as Bibles. Hemp paper offered major advantages in that it could be recycled up to eight times versus three times for standard pulp paper. Both Carver and Ford argued that using hemp paper would save most of America's great forests from the ax. The "waste" seeds of hemp crops could be used for animal feed, oil production, and the making of paints.

The chemurgical movement revived the interest in hemp. Hemp blended well with other fibers and could be used in papermaking, rugs, fiber-based plastics, and ropes. Many chemurgists saw hemp as Ford saw soybeans and Carver saw peanuts. William Hale, Henry Ford, Rudolph Diesel, and George Washington Carver had suggested hemp for the production of alcohol prior to World War I. Acreage doubled in hemp from 1934 to 1937 as applications grew. Ford was producing hemp injected with soybean plastic resin to make car bodies that were much stronger than steel. Just as chemurgists were opening new doors for hemp, a new problem arose in its relationship to marijuana. Hemp was illegal in most states. Ford had his experimental crops of hemp under twenty-four-hour guard. In 1937, however, the talk was to open up industrial production because of its potential. Ford was talking to Carver on the use of sweet potato resins with hemp fibers for parts. He had begun to put acreage in hemp for future production needs. He envisioned hemp being used for the production of alcohol which he had experimented with prior to World War I. He had started to advertise it with the production of a hemp car at Dearborn. Many argue that hemp's demise was at the hands of William Randolph Hearst and the DuPont Company who feared competition to their paper monopoly.[3]

The real excitement at this 1937 chemurgy conference was the prediction that, in paper production, one acre of hemp could save five acres of American forest. In addition, hemp could be harvested every four months. William Randolph Hearst had a monopoly on paper production via the pulp process and his timberlands. The pulp paper process was cheaper but the paper quality much lower, and the process put toxins into the environment. Chemurgists in the early 1930s had unlocked a new process to add hemp cheaply to paper production. Hearst had a big ally in DuPont who owned the pulp sulfide process. Hearst would deny it, but he used control of the press to launch a yellow journalism campaign against hemp with scare tactics about its potential drug use. The fear was compounded by the increase of drug use during the Depression. DuPont Company joined in the political fight since they owned the patent for nylon, which was a direct competitor with hemp as a fiber. The same year of 1937 that Carver and Ford met in Dearborn, the U.S. government banned the growing of hemp. During the 1940s' war years, Carver suggested a method of separating the toxic drug from the fiber production, which would allow for long-term legal production of one of nature's strongest fibers.[4] Carver's discovery was overlooked as Big Chemical was investing in the production of nylon.

Another topic of this historic conference in 1937 was that of flax production. Flax was another strong fiber that was a favorite of Henry Ford and Carver. The conference speaker that day was Harry Hans Straus who had spoken earlier on the government shutdown of his hemp papermaking operations. In 1937, he had turned to making fine paper from flax. Prior to World War I, Austria controlled the manufacture of high-quality lightweight paper using flax and hemp. It was the fine paper in Bibles and cigarette paper. Seventy-five percent of American cigarette papers came from Europe. With the war, Harry Straus was making it in France, but he planned to move production to America in 1937. He had invented a separator to take out the woody portion of flax since hemp was illegal. Straus was enlisting the help of chemurgists at the conference to help expand flax paper production. Like other chemurgical products, he faced tough resistance from the cotton lobby, New Dealers, and the Roosevelt administration.

Straus did get the help and started the research needed to produce flax paper in America. Straus would exchange hours of discussion with both Ford and Carver at the conference on the potential of flax. Carver had been a supporter of flax going back to the days of his youth where he was familiar with flax production in Missouri. Carver had suggested it on his arrival at Tuskegee as a substitute for cotton. Carver had even grown flax at his experimental station at Tuskegee, but when he asked for support from the Department of Agriculture, he was turned down in the 1920s. He had experimented with flaxseed oil as a substitute for linseed oil in paints because of its drying properties. Ford had also experimented with flaxseed oil in paints. Ford, of course, was already using flax for fabric in his cars. He had been using flax linen as a base for the production of artificial leather since the 1920s. Already looking to reduce transportation costs, Ford could grow flax in the North close to his factories. In addition, Ford's agricultural engineers and chemurgists in the 1920s had developed new farming and processing methods to reduce the labor-intensive nature of flax production. Ford engineers had developed new machinery to process the plant into "line flax" and "tow," reducing costs and improving quality.

In 1926 Ford had published much on the topic of flax use, predicting,

> We regard this work in flax as among the most important experiments which we are carrying on, for not only will it result in a better product than we have as yet been able to turn out, but also it will be another money crop for the farmer. We alone shall require the product of about fifty thousand acres annually, and flax fits very nicely into the rotation of crops. Thus we shall have a cash crop for the farmer and perhaps a new industry for a country. And this is not counting the value of flax by-products — linseed oil, or tow, which make excellent stuffing for upholstery.... It could be made by village industry manned by farmers, who can apportion their time between farm and equipment.[5]

Straus worked to develop a short-stem variety of flax for paper production after the 1937 conference. By 1940, Straus was reporting 300 pounds of linen fiber per acre versus 200 pounds for cotton. Straus continued to expand the use of flax as America entered the war, and he developed new products such as rope, twine, shoestrings, and parachute harnesses. Flax fiber factories also produced linseed oil, and waste went into wallboard, plastic panels, and linoleum. Carver had predicted many of these pressed products with his earlier use of peanut shells in them. Henry Ford had also pioneered the use of flax waste in plastic car parts.

Flax and its resultant fiber linen had a long history going back to the prehistoric cave dwellers and Egyptians using it to wrap mummies. Of course, Henry Ford's Ireland had been making linen since 500 B.C. by the earliest colonists who brought flax to grow and spun their own home fiber. It was extremely popular in Ireland for clothing because it grew well in the northern climate. Flax and hemp for cloth had been popular in states like Missouri and Kentucky. Carver's family farm raised both flax and hemp. His adopted mother made both fiber and clothes from flax and hemp.

Slaves in the South were also adept at making their clothes from flax linen. Cotton, which was picked by slaves, was sent in raw forms to the factories of England and New England, only to be imported back to the South as cloth. Slaves could not afford cotton shirts; thus, they often wore home grown and spun flax linen. Similarly, the poor of Ireland had made their clothing from refuse flax. Interestingly, the ancestors of both Ford and Carver had worn flax shirts. For Booker T. Washington, the linen shirt was more a symbol of slavery than cotton.

Washington noted in his autobiography, *Up from Slavery*, "The most trying ordeal that I was forced to endure as a slave boy was the wearing of a flax shirt. In the portion of Virginia where I lived it was common to use flax as part of the clothing for slaves. That part of the flax from which our clothing was made was largely refuse, which, of course, was the cheapest and roughest part. I can scarcely imagine any torture, except, perhaps the pulling of a tooth which is equal to that caused by putting on a new flax shirt for the first time."[6] Ford's chemurgists had learned to process various grades of cloth from flax with the roughest fibers (the "tow" of Washington's shirt), reinforcing strong plastic parts and artificial leather, while using a better grade for upholstery. In addition, Carver and other chemurgists had developed longer-strand plants.

Amazingly, as Carver noted in 1937, America was importing more flaxseed and oil than it produced. Linseed oil of the flax plant was popular in paints and had been one of the first motivations of Henry Ford in setting up flax farms in the 1920s. Ford was probably first introduced to linseed oil as the paint base for the traditional red barn. Farmers also used linseed oil to finish their gun stocks. Henry Ford was the country's major user of linseed oil in his paints for automobiles. Linseed oil usually required a thinner such as turpentine, but dried to a hard varnish-like surface. Ford would combine his paints with those of Carver in Ford's Richmond Hill, Georgia, experimental station, which was the location of the next meeting of the two men in March of 1938.

This 1938 meeting of the two men would be a major exchange of ideas. Both men had been followers of the great German chemists and their use of fertilizer, but approached it differently. Ford was a believer in commercial fertilizer, while Carver believed in natural composting. During the visit, Carver was able to quickly win Ford over to composting at Ford's Georgia plantation. The idea of saving and recycling by composting appealed to Ford's conservation roots.

Carver and Ford grew as friends, united by a belief in industry and agriculture in the New South. The South had become Henry Ford's adopted region with Ford wintering there every year. After Edison's death, he rarely wintered or vacationed in Florida, and in the 1930s he went to his mansion, Richmond Hill, at Ways, Georgia (named after the Ways Station — an Atlantic Coast Line Railroad stop). Ford would have the name of the community officially changed from Ways to Richmond Hill in 1941. Few citizens would have cared even if Ford had called it Ford Plantation because of the economic rewards he heaped on these poor counties. Originally, Ford had heard of this part of Georgia by vagabond and environmentalist John Burroughs. The Fords had stopped at Ways Station in 1921 to first visit Berry Schools, which the Fords stayed at often when returning from Florida. Ford had purchased the old rice plantations of Strathy Hall, Hermitage, Cherry Hill, and Richmond Hill in 1924 for agricultural experimentation. These plantations covered about 100 acres on the Great Ogeechee River, about seventeen miles south of Savannah, at a cost of $2,100. The area was known as the "Black Ankle" because of swamps and mud. Most of the plantations and commercial buildings had been destroyed by General Sherman's march to the sea. The area was poor, with decaying old plantations and cabins. Its remoteness and inaccessibility made the area popular with moonshiners.

Over the next ten years Ford continued to purchase land and finally owned over 70,000 acres (100 square miles). He developed the area slowly as bigger projects occupied him in the 1920s. He would eventually spend nearly $5 million in the area. In 1925, Ford faced a major development project to make the area livable and profitable. Ford was said to have rooted out 250 stills in the area to start the cleanup.[7] In addition, Ford drained the swamps which were a source of malaria. The sole population of the area was poor black farmers and moonshiners. He built two medical clinics to serve the area and staffed them with nineteen nurses. Ford offered free immunization against typhoid, smallpox, and diphtheria. In schools, Ford arranged for dental work for students as well as free glasses. Every child was given a health exam every year. Once he had addressed the poverty, he looked to economic development. Since the area suffered from massive unemployment, Ford went about employing all in work, farming, teaching, and craft production.

The first phase of farm development was to convert the rice fields to other crops. Ford started to restore the sawmill and utilize the trees from land clearing in building. He also rebuilt the brick-making kilns to produce building brick. Ford started to rebuild the community with these resources, adding schools, churches, and community centers. Then Ford moved to make use of the land. This was achieved by planting lettuce, which Ford sold for a profit over the next ten years. He even built an ice plant to sell ice to the railroads to ship his lettuce. Early on, Ford also cleared new land for an experimental farm.

Ford, Edison, and Firestone had formed the Edison Botanic Research Company to research a domestic source of rubber. The Fort Myers, Florida, lab was too crowded for a farm, so Ford hoped to use his Richmond Hill plantation for Edison to test new rubber plants. The area was planted in experimental goldenrod which Edison believed could produce enough latex to make good rubber. Edison's death in 1931 forced Ford to bring Harry Ukkelberg, a former Edison employee at Fort Myers, to Richmond Hill to take care of the experimental rubber operation. By 1937, the German synthetic rubber process and Firestone's own investment in Liberian rubber plantations pushed Firestone out of the partnership. Ford's goldenrod rubber was still at two dollars a pound versus synthetic rubber at thirty cents a pound, while plantation natural rubber was at twenty-two cents a pound. After Ford's discussions with George Washington Carver in 1937 in Dearborn, Ford changed his direction at Richmond Hill from mainly rubber to add new areas. Ford looked to these areas at Richmond Hill:

1. More uses for farm waste in the line of Carver's extensive work in composting and fertilizer production. In addition, study the uses of things like corncobs, pine bark, sawdust, and used clay

2. Ford also wanted to study the production of ramie grass fiber, one of the strongest grass fibers that Carver had experimented with. Ford also wanted to study bamboo as a fiber for paper.

3. Expand sweet potato production and do experiments with growing safflower and perilla.

Ford also involved Carver in following up on Edison's work on rubber. Ford funded a new laboratory for Ukkelberg and hired twenty "assistants" at Richmond Hill. Actually, these so-called assistants were unemployed moonshiners who were on probation. Experiments included over sixty varieties of goldenrod for improved rubber yield. Unfortunately, Carver's health was declining and prevented him from fully participating in the Richmond Hill research.

Ford had another purpose in hiring moonshiners; he wanted to explore new possibilities in alcohol production. This was part of Ford's chemurgy interest in an alcohol/gasoline mix to fuel autos. Ford actually started looking into cellulose sources for alcohol,

13. The Meeting of Ford and Carver

Carver in his museum, 1937 (Alabama Department of Archives and History, Montgomery, Photograph Q4843).

which is a popular research area today. Like Ford, Carver had an interest in wood alcohol as a fuel source. In the late 1930s, George Washington Carver and his assistant, Austin Curtis, made a numbers of trips to the Richmond Hill laboratory to work on these projects. The laboratory was particularly focused on composite materials, where Carver had extensive knowledge. The laboratory had found a method to produce plastic tiles from corncobs, albeit very expensively. Ukkelberg even developed a method to make rayon out of sawdust and delighted Ford by sending him a pair of rayon socks.[8]

Starting in 1935, Clara Ford became involved in the building of a winter home at Richmond Hill. Both she and Henry would spend much time living out of their private railroad car, *Fair Lane*, while designing this new mansion. The mansion was to be a Greek Revival structure of epic proportions. Ford purchased a number of decaying antebellum mansions for bricks in the new mansion. Many bricks came from the nearby old Hermitage plantation home. The slave quarters of the Hermitage were sent to Greenfield Village as an exhibit on the task system on slave plantations. In addition, Ford started the rebuilding of the whole community with over 100 homes. The mansion known as "Richmond" would include a huge ballroom for dancing. Ford often brought in musicians from Detroit for community dances. He added his own power plant as he had in Dearborn to supply air-conditioning (a rarity at the time). Also like Dearborn, Ford built a private laboratory for himself and friends, such as Carver. Quarters were prepared for Carver and assistant Curtis to stay in when they visited. Clara and Henry would spend three months of winter there every year.

Construction in the community continued into the late 1940s. Ford focused on improving the area's educational system and schools. He took over the rebuilding of seven one-room schoolhouses for both blacks and whites. Ford brought in teachers and demanded his McGuffey-style approach. At night the schools ran free educational programs for illiterate field hands and workers. In 1939, Ford built a larger consolidated school named the George Washington Carver School, which was dedicated by Carver. This school would eventually have all the grades through high school and would function as a vocational school. The educational approach included that of nature study and gardening designed by Carver. Carver and Ford had home economics added to the curriculum. Ford added vocational training including welding, blacksmithing, and metalworking. Like his series of schools in Michigan, he included mechanical farming and machinery repair. Ford provided bus service, books, equipment, libraries, a chemistry lab, a blacksmith shop, and cafeterias (lunches were provided free). As in Dearborn, Ford built a Martha-Mary Chapel, which he had prefabricated at his plant in Massachusetts, and held daily nondenominational services for the students.

14

The New South, Soybeans, and Final Dreams

A vision of a New South had been part of Carver's mission, and in the late 1930s, he fused this vision with Henry Ford's. Ford had also envisioned a New South in the 1920s with his failed Muscle Shoals venture. Both men saw poor agricultural management and lack of supporting industry as the root cause of the depressed South. The South still reflected the plantation system of centuries earlier. The Civil War broke the King Cotton hold on a people, but cotton still enslaved the region economically. The problem was deeper than just cotton itself. Both Carver and Ford realized that the soil of the South was inferior to the rich mineral dust of the Midwest. Professor Henry C. Wallace had trained Carver that a measure of a nation was its topsoil. At a talk at Henry Ford's Ways, Georgia, plantation in 1938, Carver noted, "We must enrich our soil every year instead of merely depleting it.... There is a distinct relationship between the soil and the person living on it. Show me a poor lot of land and I'll show you a poor farmer."[1] Ford had proposed vast fertilizer manufacturing in Tennessee, Alabama, and Georgia, but Carver looked to composting. Both men had their differences in methodology, but they agreed on the huge potential in the South through the proper enriching of the soil.

By the late 1920s, there was talk of a Southern Renaissance based on the new combination of industry and agriculture. Government approaches had all failed; Carver and Ford believed the renaissance would come from its people. Carver favored a new crop strategy, small farms, and local industry. Ford looked to more specialized crop production tied to village industries. Carver favored a farm-centered strategy, while Ford favored a farm-village strategy. These were not major differences but different paths to the same end. Carver had preached sweet potato and peanut crops as a soil improver, food provider, and a base for industrial products. Ford leaned more toward soybeans, hemp, and flax to improve soils and supply industrial production. Both men had faced the stubborn resistance of southern farmers to leave their cotton and the government's stubborn belief in paying farmers not to produce.

In the late 1920s, Carver's ideas on peanuts and sweet potatoes were taking root. Peanuts were becoming a commercial crop. His work with sweet potato starch had finally gained momentum with the chemurgical movement. America's appetite for and commercial uses of starches had risen to over 1,400,000 pounds per year, half of which was imported. Carver had been pushing sweet potato starch for a decade. He had planted the seed of a new southern economy. He had worked with the government on sweet potato starch in the shortages of World War I. Carver had envisioned a substitute of cotton acreage to sweet potatoes and peanuts. In 1928, the country's first commercial sweet potato starch plant began in Louisiana, but it struggled as a food stock due to public resistance to nonwhite starch. Carver, however, convinced the government that there was a future for sweet potato starch in food, textile mills, laundries, confectionary, brewing, and commercial alcohol production.

Sweet potato starch became popular again in the 1930s, and the government funded research in the manufacture of white sweet potato starch based on the early work of Carver. The South finally realized Carver's vision in sweet potatoes, which had up to three times the starch of corn. The Federal Emergency Relief Administration allotted $150,000 to build a sweet potato processing plant at Laurel, Mississippi. The Laurel plant was in an area that formerly was totally dependent on cotton. Experimental stations started work on sweet potato and yam breeding for more starch and less color. Breeding further improved disease resistance and yield, which increased from 70 bushels an acre to 600 bushels.

Carver continued his endless promotion of the sweet potato as an alternative crop to cotton. He expanded into its cousin, yams, aware that slaves going back to the 1700s had eaten yams. Carver had fully researched slave foods over the years. The African yam (similar to the sweet potato) was often used as a staple on slave ships. Similarly, slaves in the colonies had eaten black-eyed peas, also known as cowpeas. They even had recipes to make a coffee out of okra. Slaves had made soup from peanuts and sesame seeds. Carver's recipes clearly drew from this culinary history to expand the food potential of these alternative crops. He was having success with the sweet potato recipes by the mid–1930s as sweet potato pie became common on southern tables during holidays. By the 1940s, sweet potatoes had taken their place on the Thanksgiving table across America.

World War II brought new opportunities and challenges for both Carver and Ford. Carver, who had helped the nation save and find food substitutes during World War I, was called on once again. Senator Lister Hill of Alabama asked Carver to review his work on using peanut oil in soap making to replace coconut oil from the Philippines. In the end, peanut oil was too expensive in soap making but was a great replacement for imported olive oil. Carver issued his next-to-last bulletin, number 43, *Nature's Garden for Victory and Peace*, which was similar to those he had issued during the First World War. This bulletin became the basis for a *Saturday Evening Post* article that popularized the "Victory Garden" concept once again. Carver also worked on two leaflets—*Food, What Is It? How Can It Win the War?* and *Peanuts to Conserve Meat*.

Carver's sweet potato glue became popular again for applications such as stamps during the war. This type of glue was starch based. Sweet potato starch once again became an important raw material during the war as imported starch was cut off. Some of these secondary sources such as tapioca starch in Florida and imported tapioca from the West Indies had been cut off. Sweet potato starch once again played a bigger role in laundering and the sizing of cotton during the war. Starch imports had been growing since 1922 to the start of World War II, from 200 million imported pounds, or 25 percent of domestic consumption, to 500 million imported pounds, or 38 percent of domestic consumption.

14. The New South, Soybeans, and Final Dreams

The difference in imports was made up mainly of sago palm and tapioca starch from the West Indies. Tapioca starch was superior to corn and white potato starch in remoistening glue in stamps. Carver's sweet potato remoistening glue became a perfect substitute as stamp glue during the war.

Carver's sweet potato starch would again save the nation by augmenting its starch resources. America was very dependent on cornstarch and imported starches at the start of World War II for a large variety of necessary products. The country was using over a billion pounds of cornstarch alone. Only 250 million pounds were being used for food. The sizing of cotton textiles used 300 million pounds, and the laundering of garments used 250 million pounds. Finally the paper industry used about 200 million pounds. Another 100 million pounds of cornstarch was used to produce dextrin for glues, adhesives, and other industrial substances. On the eve of war nearly half of this starch was being imported. With the war increasing demand and reducing imports, government chemists and private chemurgists were again studying Carver's sweet potato research.

Another area where Carver and Ford had prefigured an important breakthrough was the production of synthetic rubber. While both Ford and Carver had favored domestic plants over chemical synthetic rubber, their early chemurgical work on grain alcohol would be a key to synthetic rubber production. Carver had been making alcohol from sweet potatoes in the 1910s prior to the giant German chemical firm IG Farben, which started to use sweet potato alcohol to make synthetic rubber in the 1930s. In fact, the Germans, like Carver, were breeding better sweet potato variants for industrial uses. The rubber was known as Buna-S or Butadiene rubber, and it could be made from petroleum or alcohol. The Germans used both methods anticipating an oil cutoff during war. In World War II, America was cut off from Asian rubber which created a strategic shortage.

The petroleum lobby had convinced the government to use petroleum to make synthetic rubber, but in 1941, they could not meet war requirements. In the end, it was America's chemurgic movement that produced Butadiene from grain alcohol, which saved the day. This work with alcohol-based butadiene rubber was headed by Reichhold Chemicals. In 1943, alcohol-based rubber accounted for 77 percent of the nation's total output of rubber. The petroleum-based process had failed to meet even the minimum requirements of a nation at war. The problem was in the design and building of petroleum-based rubber plants. Still, by the end of the war the petroleum produced a far cheaper rubber and became the preferred feedstock. Interestingly, with the price of oil, alcohol-based rubber is again being looked at today.

Carver, Ford, and the chemurgists realized that education would be needed to expand alternative crops permanently. Many of Carver's bulletins were once again published to educate the public on gardening and food preparation. In the late 1930s, the peanut became a major southern crop because of twenty years of education by George Washington Carver. The Cotton Belt had come a long way — Alabama had a three-day peanut festival every year, with Carver being escorted in large parades. By 1940, the peanut had become the largest southern cash crop, second only to cotton. This, more than anything, was a tribute to the humble chemist of Tuskegee. In thirty years, Carver had created an agricultural revolution in the Cotton Belt through continuous and endless education. Henry Ford hoped Carver could help do the same thing in the North with soybeans.

Carver was never in good health after meeting Henry Ford in 1937, but he was able to make a month-long visit to a specially prepared lab for him at Ford's Dearborn compound. Originally, Ford had hoped that Carver would stay on, but Carver's health was too poor. Carver did send his assistant, Austin Curtis, to work at the Nutritional Research Center Ford had built for him, and they did have some great childhood moments in the lab together. Ford also brought Carver up to date on his research on soybeans, plastics, and the ultimate production of a green car. The use of hemp and flax as base fibers in composite plastic materials was also experimented with, and Carver would return to his lab to continue the joint research. Carver promised Ford that he would move into soybean research at Tuskegee. Ford also started building a laboratory and cottage for Carver to use at his Richmond Hill plantation.

Carver and his students often visited Ford's summer home and experimental farm at Richmond Hill where Carver converted Ford to more organic methods. During the war, Ford restarted his research into domestic sources of latex for rubber with Carver's help. Ford helped rebuild Carver's laboratory at Tuskegee and focused Carver's attention more toward nutrition and medicine in his last years. Carver explored a number of products such as persimmons. Persimmons have astringent properties that Carver used to develop some dental products. Japanese and Chinese had been using these properties for years. Carver suffered from pyorrhea and was hoping to treat it with persimmons. Austin Curtis aided in the preparation of extracts, and both men appeared to have some success with local dentists. Still, nothing ever came of it commercially. They both continued their work with soybeans and explored Osage oranges ("horse apples" in the Midwest; "monkey balls" in Pennsylvania) for medical and nutritional uses. Actually, Carver had started his work on Osage oranges as a potential source for rubber at the suggestion of Henry Ford. Edison had looked at the Osage orange, too, while at his Fort Myers lab, and some were planted at the Richmond Hill experimental station. Rubber research once again became a hot topic with the war in the 1940s, and the experimental fields were already at Richmond Hill. In addition, Carver and Curtis traveled to Ford's experimental Richmond Hill to consult on fertilizers, farming, alternative rubber, and other Ford interests.

More than anyone, Ford helped Carver receive the recognition he deserved. Carver's success with education, peanut and sweet potato products, and race relations had brought him national attention, but honors had been slow in coming. In 1923, he won the Spingarn Medal of the National Association for the Advancement of Colored People. Later in 1928, he received an honorary doctorate degree from Simpson College. The doctorate was his most cherished award. He received the Roosevelt Medal from the memorial society of Theodore Roosevelt in 1939. It was presented to him in New York where he shared the honor with Carl Sandberg. These were but a few of his awards prior to 1940 that established him as a scientist of note. Invitations poured in for Carver to appear or speak, but his health was declining in the late 1930s, preventing him from doing so. However, Carver could not resist the chemurgical movement which was in line with his years of research

14. The New South, Soybeans, and Final Dreams

Carver in 1937 at the reconstructed home cabin at Greenfield Village (Bentley Historical Library, University of Michigan, File HS8577).

in economic biology. In the 1940s, Carver received a second honorary doctorate from the University of Rochester.

Carver's greatest honors, however, would come from Henry Ford rather than from his degrees. Ford helped Carver gain the recognition as a scientist that he truly deserved, and Carver himself acknowledged this in a letter to Ford in 1941.[2] Ford would name a school after Carver in Ways, Georgia, in 1940; and in 1941, Henry and Clara came to Tuskegee to dedicate the opening of his museum. Ford and Clara inscribed their names in the cornerstone and brought a number of soybean car parts to be included. Plans were set for Carver to come to Dearborn to dedicate a Carver laboratory for research and Carver's cabin in Greenfield Village. Assistant Austin Curtis had been in Dearborn working on both projects. The Greenfield Village project was to build a replica of Carver's birthplace.

The Carver cabin would have great symbolism, which Ford loved. In the village, the cabin was placed between the Hermitage slaves quarters and Lincoln's Logan County Courthouse. It was a symbol of Carver's (and the nation's blacks') rise from slavery to leadership. The Carver Memorial Cabin was built of pine logs from Iron Mountain, Michigan, that were made into planks by students in the village. In addition, special representative logs were brought from every state of the union to make inside paneling. Wood panels included shortleaf pine from Alabama, apple from Michigan, redwood from California, buckeye from Ohio, and holly from Delaware. Carver would spend a night there as part of the dedication in July of 1942.

The final tribute by Ford was the Carver Laboratory, which Carver dedicated and spent several weeks at in July of 1942. A number of dignitaries were there at its opening, including Charles Lindberg. The laboratory was on Michigan Avenue and was named the "Nutritional Laboratory of the Ford Motor Company" (changed to Carver National on Carver's death). The laboratory had been an old water plant in Dearborn, so it had extensive underground storage for fruits and vegetables. Ford built a greenhouse for experimentation and had an enclosed, fully equipped chemical laboratory. Ford added a well-stocked library in the style of Edison's laboratories. More to Carver's style, an experimental kitchen was added. For the opening, two types of "weed" sandwiches were prepared and recipes given to the press. One spread was a mixture of more traditional vegetables such as carrot, endive, celery, onion, kale, and others with a little ham. The other spread was dandelion, chickweed, lamb's quarters, plantain, bergamot, oxalis, and other wild plants. Of course, Carver and Ford chose the "weed" spread for lunch.[3] In addition, there was soybean ice cream and milk.

Henry Ford had taken great care to maximize the publicity of what amounted to a two-week visit in 1942. Ford talked again of Carver working on making rubber from sweet potato, dandelion, and the old Edison rubber recipe from milkweed. Henry Ford had employees using a vacuum cleaner to gather Dearborn-area dandelion seeds for Carver to study as rubber sources. Of course, soybeans were always central to Ford's farm research. Ford truly believed that soybeans would replace wheat and corn as America's crop, but Ford had a lot of convincing to do. He hoped that Carver could create the same magic for soybeans that he had for peanuts. Carver's health would, however, be the roadblock. But for a couple of weeks, the old lions dreamed new dreams.

While Carver's visions were realized in the South, Ford was seeing his stirring in the North. In the 1930s, soybeans awakened as a food product. For years, Ford employees had enjoyed soybean cookies and bread, but few found interest in Henry Ford's diet, with a few exceptions such as Carver, who had first recognized their food value. World War II created huge demand for soy oil as imports were cut off. Production for oil skyrocketed and created a huge surplus of soy meal. Ford's Dearborn lab turned to researching the use of soy meal. Carver and his assistant were well suited to help in this area. Soy meal badly needed new applications as the peanut had decades earlier.

The chemurgical movement had pushed soybeans in food and industry. With the chemurgical movement, American universities in the Midwest created their own research into soybeans to help local farmers. From 1936 to 1941, soybean production multiplied seventeen times, approaching the subsidy which boosted the big four: cotton, wheat, corn, and tobacco. In 1942, the United States overtook Japan to become the world's second-largest producer behind China. Soybean production was based in the Midwest, but finally the South starting producing it for animal feed and soil improvement as Carver had suggested decades earlier. As chemurgical applications increased, both Ford and Carver

14. The New South, Soybeans, and Final Dreams

became legendary in the movement for their early visionary work. Unfortunately, Carver's health problems prevented him from making these breakthroughs. However, Carver's assistant, Austin Curtis, went to Dearborn to work in the lab for a year.

While physically limited, Carver continued to exchange letters and make visits to Ford's Georgia laboratory. Carver experimented with soybeans, peanuts, sweet potatoes, and pecans to develop new foods. He created a cake covering and filling of sweet potatoes and soybeans. Carver sent Ford one of his cakes, and Ford suggested the possibility of creating a breakfast food. Carver also developed a salad recipe of peanut and soybean sprouts for Ford. In 1942, Carver exchanged letters, recipes, and food products with the scientists at the Dearborn laboratory. New cosmetic products for soybeans were created by Austin Curtis.

In 1943, Ford would bring Holton Diamond from Ohio to head up the Carver lab in Dearborn. Diamond had a degree in chemistry from Wilmington College and continued to work on his master's at Wayne State University. Diamond advanced the work in developing nondairy soybean products such as whipped topping, coffee creamer, and ice cream. Diamond's patents would lay the groundwork for products we commonly see in the nondairy section today. Diamond was very successful, but Ford's death in 1946 started the closing of the laboratory. Unfortunately, the laboratory and Ford's chemurgic projects would not be part of his legacy.

One project of Ford's that would make up his legacy was the Henry Ford Museum and Greenfield Village. In the 1930s, Ford expanded on its theme of American exceptionalism in industry and agriculture. It was to be an inspirational hands-on museum for America. Ford gets little credit for the genius of this museum. He added more from his and Edison's industrial beginnings, including a reconstructed Detroit Edison Company's power plant where he had been chief engineer. He reconstructed the Bagley Avenue watch shop of Robert Magill, where as a sixteen-year-old he had repaired watches. He also reconstructed the Sims and Armington Machine Shop of his youth. He reconstructed his first simple factory with the Old Owl Lunch Wagon, where he often ate his dinners. He built a crafts village with a pottery, glassworks, blacksmith shop, cooper shop, and textile mill. He even reconstructed the Plymouth carding mill, where he had taken his father's wool to be carded for spinning. He bought and reconstructed an early water-powered silk mill from Connecticut. Ford, always the farmer, even brought silkworms and mulberry trees to Greenfield to try his hand at silk production.

Ford implemented a theme of showing industry's roots in the early American crafts evolving into factories. He brought the Wright brothers' shop to Greenfield Village in 1937. He also added to the indoor museum with an endless array of products showing this evolution over the years. Another theme was consistent with that of Carver, McGuffey, Edison, and others. This was the fusion of practical skills with scholarship in education. Ford added to the village the homes (or replicas) of Noah Webster, Edgar Allan Poe, Walt Whitman, and Stephen Foster. He added the house where Robert Frost lived while in Ann Arbor at the University of Michigan. To fill in his village of Americana, he added stores, inns, stagecoach stops, post offices, the courthouse where Lincoln practiced law, and an imbedded school and research center.

Besides finishing Greenfield Village, Ford was concluding work on his last engineering dream — the plastic Model T. The soybean lab at the village had been at the heart of this effort. On August 13, 1941, at a huge luncheon, Ford unwrapped a new "soybean" car. Lunch was a fourteen-course soybean-based lunch for the press reporters and engineers where Ford sported his soybean "wool" suit and tie (actually 25 percent soybean, 75 percent wool). The car was an all-plastic soybean body and had many plastic parts. Using a hammer, Ford showed its dent-resistant properties over steel, which is what makes tough plastic car hoods and doors popular today. Ford said molded plastic parts were actually cheaper. The molded side panels were 70 percent cellulose (mostly hemp) and 30 percent soybean resin. The steering wheel was made of hemp fiber and soybean resin. The upholstery was soybean "wool." The car shaved 1,000 pounds off its steel counterpart, offering considerable gas savings for the consumer. In addition, it was a fully green car running off grain alcohol. World War II would prevent Ford from bringing his dream of mass-produced soybean cars to reality. At the same time, Ford announced his opening of a pilot plant to produce soybean fiber as a replacement for wool. Ford predicted he would cut his upholstery costs for cars by 50 percent. During the World War, Ford's soybean wool replaced 30 percent of the imported supply of wool that was cut off.

While Ford was finishing his work on Greenfield and the soybean car, Carver was working on his legacy at Tuskegee. He continued to work on his own museum. In 1940, Carver used his lifelong savings of $60,000 and established the Carver Foundation at Tuskegee Institute. It was an amazing sum of money, which was a tribute to his lean living. The foundation was to continue his research and to build a museum. No doubt he was influenced by Henry Ford in this effort to make a continuing statement about his beliefs and work. The Carver museum, albeit on a much smaller scale, had the same mix of education, hands-on interaction, inspiration, and personal history.

15

Home and Industrial Economics — Lean Manufacture and Household Savings

When the nation faced shortages during two world wars and the Great Depression, the nation turned to Ford and Carver. Ford knew how to run a factory, and Carver knew how to efficiently run a household and a small farm. Ford turned his factories into massive producers of tanks and airplanes at a rate faster than any enemy could destroy. Ford ran his organizations lean with recycling and remanufacturing. He showed flexibility in using America's agriculture in ways beyond food production. On the other hand, Carver could teach a household to survive on little, and to farmers, ways to increase yields. Both men argued that American farmers were a limitless resource for industries and families. The waste of both industry and American households was appalling to both men. Thank God they never lived to see a garage dump today. Still, these men showed the way in turning this waste into surplus.

Carver and Ford were like souls from different cultures. They had a passion for saving and efficiency, and recycling was a passion for both. They saved and recycled at home and work. Carver was a library of household saving tips and formulas. When vegetable oils were too expensive for homemade paints for poor farmers, he developed paints using used motor oil, a waste product. Carver developed a filtering process using activated carbon and kaolin to take out carbon and metal particles in the oil followed by a sulfuric acid treatment to further clean it. In the 1930s, he was awarded a grant from the government to further develop this process with his assistant, Austin Curtis. The production of low-cost paints proved to be a major saving on government work projects of the period.[1]

Carver recycled almost everything for fertilizer including human waste. He preached that any street sweepings, sawdust, chips, and rags should go to the compost pile. Similarly, Ford recycled wood, paper, and cardboard at his factories. It was the same at the household level for Carver; paper was pressed into fireplace fuel or wall insulation. Carver developed solvents and bleaches to remove ink from paper to reuse it much like Ford cleaned and even recycled rags in his factories. Caver remembered his years on the prairie where he learned subsistence living out of necessity. He abhorred the waste of fruit in the South and developed drying practices to preserve fruit for the winter season. He applied these drying practices to sweet potatoes as well and also developed a potato chip product for storage. Carver had also envisioned the lean shipping of dried fruits in the 1920s, since up to 90 percent of the weight was water. The saving would be enormous by shipping dried product. Freeze-dried food today is similar to Carver's concept in the role of dried food.

Nuts as well as fruit commonly rotted on the trees in the South. Carver preached the use of natural nuts for human consumption and for animal feed, including the common acorn. Walnuts, beechnut, pecans, and hickory nuts could all be made into nut cakes. He argued that southern farmers with a small lot could live off sweet potatoes and peanuts supplemented by wild fruits, nuts, and weeds much like the prairie subsistence living of his youth. Carver identified many common "weeds" that could be used as food, publishing a separate bulletin on this. He proposed the new uses of pecans which could be readily grown in the South. Carver listed a number of methods, tips, and plants that would allow the southern farmer to live off the land. He published a leaflet, *Are We Starving in the Midst of Plenty? If So Why?*, during World War II. Carver showed farmers how to live free off the land through Tuskegee seminars as well. Many a black family did just that during the world wars and the Great Depression.

Waste at any level seemed to motivate both Carver and Ford. During one trip, Carver looked at the industrial waste around the West Tulsa Refinery of the Mid-Continent Petroleum Corporation. The refinery had a field of sludge and heavy residuals, not uncommon in those days. Carver took samples of the waste refining residue back to Tuskegee and, not surprisingly, found some potential uses. The heavy sludge could be formed into asphalt and a crude rubbery product. Carver was able to extract dyes from the black sludge as well. This was not unlike Henry Ford in his early studies of coke production at the River Rouge plant where he developed an array of commercial by-products from what was previously considered industrial waste. There was no part of the massive waste from the coking process that Ford didn't use. He would eventually find that coke by-products were almost as valuable as the coke itself, opening the eyes of steel producers who for decades had been making coke and dumping these valuable "waste" products.

Ford and Carver had a propensity to make the elimination of waste a religion. Furthermore, Ford liked preaching to industry, and Carver often preached to the farmer on waste. In a 1936 letter to *Peanut Journal*, Carver noted, "Now is a crucial time to chemicalize the farm. We must not only make the farm support itself, but others as well, with a large manufactured surplus to sell to those who are fortunate enough to own and properly care for the farm.... Insulating boards, paints, dyes, industrial alcohol, plastics of various kinds, rugs, mats, and cloth from fiber plants, oils, gums, and waxes, etc., all or much of it can be made from waste products of the farm."[2] For Ford and Carver, waste was the real issue even behind unemployment. Decades after Ford brought about the climax of the first industrial revolution through his gospel of waste, Toyota would use the same principles of waste elimination to launch a second revolution known as lean manufacture.

Lean distribution was another point that both Carver and Ford agreed on. Ford had always argued for energy to be produced near the mines versus shipping coal. Both men found it extremely wasteful that agricultural products were shipped long distances to places where local supplies could be developed. It was true of manufacturing too; it should be as close to the market as possible to reduce costs. Carver argued for factories in the

South to process peanuts and sweet potatoes, and Ford built his factories in the regions and countries where there was a market and ample resources. Ford's dream of Muscle Shoals was to be a seventy-mile stretch of localized production of everything from energy and food to finished products.

Contrary to Ford's and Carver's ideas, the nation's industrial and agricultural production was spreading out. Of course, both men argued in times of abundant energy, when such waste was accepted and tolerated. In times of war, however, victory gardens and local manufacture made sense. Carver and Ford both had a commonsense approach to the superiority of local manufacture in all industries. In 1927, Carver gave a lecture at the State Negro Fair in Tulsa, Oklahoma. Before the lecture, he visited a local drugstore in Stand Pipe Hill and then collected some local plants to illustrate the point. In his lecture he noted: "I found down at Ferguson's drugstore seven patent medicines containing in their formulas certain elements also contained in these plants on Stand Pipe Hill. The preparations were shipped in from New York. They should be shipped out from Stand Pipe."[3] Carver further argued to his fellow chemurgists that the South had all the resources for an economic renaissance based on chemistry. Both Carver and Ford understood that efficient capitalism is local; distances from the market exchange only add costs. Ford, in particular, had made billions by using cost accounting at the plant level; but on a global view, common sense always trumped accounting. This is why Ford always built factories in the countries he wanted to supply.

One area that both men saw as extremely wasteful was the importation of food and agricultural products into the United States. Both men looked at the overall energy balance of shipping in products or substitutes that could be grown or produced locally. Ford had always understood that shipping American-built cars to Europe was wasteful. Building them there and using local labor and resources made the most sense. Ford was a commonsense manufacturer, and he understood that the importation of heavy manufactured goods cost more regardless of the artificial price that may be set due to currency fluctuation, currency manipulation, government intervention, or, even worse, banker manipulation. Ford built manufacturing plants and supply chains, and he knew the real costs were based on common sense. Common sense said you couldn't supply rubber from Asia, steel from the South, linen from Europe, electricity from Niagara Falls, oil from the Middle East, and coal from Alabama to make a car in Detroit and then ship it to Ireland to sell at a lower price than a car made locally. Local domestic manufacture was always cheaper from a production standpoint, and any price advantage of imports was usually artificial. Ford had spent his life shortening the length of the supply chain and the wasted energy of moving materials and products. Furthermore, Ford argued that domestic production and supply always meant independence for the company and the country.

Similarly, nothing bothered Carver more than to have a meal with a local farmer in Alabama with California and Florida fruits, potatoes from the Midwest, milk from Pennsylvania, cheese from Wisconsin, beef from Texas, and pork from Illinois. Carver argued with the stupidity of 25 percent of the nation's peanuts being grown in Alabama and then shipped to New York to be processed when processing plants could be built in Alabama. Like Ford, Carver argued against the logic of bankers, which made "sense" out of shipping products thousands of miles. It was why New York bankers like J. P. Morgan never supported the tariff policies of the Republican Party. Banks made more money from ships, railroads, and transportation than manufacturing.

Carver understood, too, that the real roots of slavery were economic. Cotton production had enslaved the South and continued to do so after the Civil War. Poor farmers sold cotton to purchase food they could have grown themselves. Tenant farmers were forced to grow cotton to pay the bills. Carver argued that true freedom is in the self-sufficiency of the black farmer. It was the very same logic that Ford had used to make his company independent. It was the heart of the chemurgical movement to make a nation free of foreign influence. Ford and Carver both saw economic freedom as freedom from war as well.

One area where their views were different was the concept of the American farm, but the root concept of freedom was the foundation. Carver believed that individual farmers should first produce food for their tables and then produce a diversified mix of cash crops. These farmers would supply local industry, a view distinct from his southern experience. Ford saw a specialized farm versus the traditional mix of dairy, poultry, and crops, but still supplying local industries. Ford's view came from the overdiversified nature of Michigan farms, not to mention his personal hatred of farm chores such as milking cows on a farm for corn. In 1926, Ford noted this vision of flax farming: "The flax growing, spinning, and weaving can and ought to be decentralized, so that it can be complementary to well-conducted farming — that is, grain farming as distinguished from dairy framing, stock farming, or truck farming. The place for the gins, the spindles, and the looms is out in the country where the flax is grown. It could be made by a village industry manned by farmers, who can apportion their time between industry and factory."[4] Of course, Ford would fully implement this approach in the 1930s. Carver would fear any dependence on a specific product coming from the Cotton Belt. Ford saw great inefficiency in the mix of cows, chickens, and crops. Carver, on the other hand, could find a type of synergy in such a mix, with animals supplying fertilizer and crops supplying feed. Ford and Carver, of course, were joined by a belief in an overall lean approach.

Ford applied his concept of time and leanness in automated production to the farm as well. He used his huge Dearborn farm to illustrate his point. In 1922, Ford made headlines by operating his farm one year without the use of a single horse. By the end of the 1920s, Ford's farms had put together a routine to do all the work of grain production in twenty days. Ford argued the fields could be plowed in two days, cultivated in five days, harvested in two days, and necessary chores such as fertilizing and fencing done in ten days. Ford called the horse a 1,200-pound hay motor of one horsepower. In fairness, Ford had an army of tractors available to do the work. It also required some specialization to eliminate the time-consuming areas such as dairy production. While the concepts of Ford and Carver on waste and efficiency were similar, it was unlikely that the smaller black southern farmer could ever achieve the scale of economies of the Ford farm.

While Ford's concept of a lean farm focused on efficiency and specialization, Carver's lean farm focused on farm economics and self-sustainability. Both the methods of Carver and Ford were centered on the elimination of waste. Carver had published bulletins during the boll weevil infestation, World War I, the farm recession of 1922, the Great Depression, and World War II on how the farmer could grow and make all that he needed to live. Carver eliminated much of the waste by recycling it, and Ford did the same in his manufacturing plants. Ford eliminated waste through specialization. Carver's approach required a type of diversity because it was based on family subsistence. Carver saw animals as a source of fertilizer and food. In the end, both men would come to middle ground. Ford, in particular, came to love composting because it was recycling. In the 1940s, both men would come to a compromise in the operation of the Ford Richmond Hill farm and plantation. However, while Richmond Hill pioneered southern farming methods and new crops, it reported losses in the 1940s.

Ford had become obsessed and deeply passionate about waste in his operations and running all operations on a lean basis. We have already seen this with the lumber-saving department at the Rouge River Plant, but Ford went far beyond that. Old paint was scraped and reclaimed for rough painting in the operations. Ford had men recycling and repairing tool handles. Recycling paint was an idea shared with Carver. At the Rouge, two men spent their time repairing mop pails. This was part of a larger salvage department that repaired tools, equipment, and belts. Ford claimed he saved $1,000 a day in salvaging belts alone. In his machining operations, he not only recycled scrap metal chips but also salvaged the machine oil off the chips! Bricks were crushed and made into new ones — something Carver had done in the early building of the Tuskegee Institute. Ford recycled twenty tons of paper a day, using this paper to make pressed board as Carver had often done at Tuskegee. Both Carver and Ford scraped old paint off buildings to make new paint. At the Rouge blast furnaces, Ford recycled slag into an array of products including cement and fertilizer. Other than a handful of Toyota plants today, none have pursued waste elimination this far. Ford summarized the importance of waste elimination in 1926: "Industry owes it to society to conserve material in every possible way. Not only for the element of cost in the manufactured article, although that is important, but mostly for the conservation of those materials whose production and transportation are laying an increasing burden on society."[5] The power of Ford's approach is its application today, and its potential to bring environmentalists and industrialists together for the future.

Clearly, this also reflects Carver's approach of waste elimination around the farm. Carver pushed the farmer to recycle everything as well. One student of Carver's would later write him that from "the savings of grease from my kitchen I have made one hundred and sixty pounds of soap."[6] Like Ford, Carver pushed recycling and waste control to extremes. As we have seen, he even used human waste in the compost pile for fertilizer. Ford would stop his car to pick up a glass bottle to recycle. This extreme and even obsessive saving is what made the approach so successful for both men. In Ford's case, it created a culture of extreme savings that pushed those less obsessed to conform to the organizational norm. Ford had men pick up old oily rags in the plant to be cleaned and reused. Carver had suggested many times the recycling of paper and cardboard for use in insulation and homemade wallboard. Carver had also gone to extremes in recycling old paint and dyes (like Ford) and using recycled oil from the tractor to make paint and grease. There was nothing that either of these two would not recycle. It was this using creativity to save ever more that made Carver so popular during the Depression and two world wars. Unfortunately, in times of abundance many forgot what Ford and Carver had taught them.

The concept of Henry Ford's lean manufacturing focused on elimination of waste and vertical integration (owning the supply chain and raw materials). Ford often chased waste into the supply chain, such as inefficient railroads, upholstery manufacturers, lumber companies, sawmills, rubber makers, rubber growers, fuel producers, and so on. Ford initially had started out as a car assembler with the Dodge Brothers, supplying major components. His highly efficient Model T plant built at Highland Park was also a basic-component assembly plant. As had other great believers in vertical integration such as Andrew Carnegie and H. J. Heinz, Henry Ford evolved into it over pricing and control issues. Ford started the River Rouge plant in 1918 to be a supplier city to his Highland Park assembly plant. The River Rouge plant took Ford into engine making, steelmaking, glassmaking, engine casting, and coke making. He employed over 75,000 employees. It took him years to move the assembly of the Model T to his River Rouge plant. In 1940, it was said that Ford at his Rouge plant could make a car from scratch in three days. Vertical integration brought down costs and increased efficiency on a scale never seen. The Rouge plant of the 1940s remains the most fully integrated factory ever built. It was the world's cathedral to vertical integration and its inherent power, but some of its secrets to success seem to have been lost with the passing of Henry Ford.

By 1926, he had vertically integrated into coal mining and transportation. A coal shortage in 1922 had Ford moving into mining. Similarly, he moved into limestone quarrying for glassmaking. His railroad tied his coal mines and limestone quarries directly to the Rouge. Ford stated about the amazing production cycle in 1926, "Our production cycle is about eighty-one hours from the mine to the finished machine [car] in the freight [railroad] car, or three days and nine hours instead of fourteen days which we use to think was record breaking. Counting the storage of iron ore in winter and various other storage of parts or equipment made necessary from time to time for one reason or another, our average production cycle will not exceed five days."[7] For Ford, time was a major part of lean manufacture, for it too could be wasted just like raw materials.

At the heart of Ford's lean manufacturing was a synergy and a spirit, much like that of Carver's frugal living. Ford's war on waste took a frantic quest. This was misunderstood by Ford's second generation of managers as well by as some old-timers such as Charles Sorensen. After Ford's retirement in 1946, many of Ford's recycling, village industries, soybean production, and community projects were questioned based on the very cost accounting system that Ford had applied in the early days of the company. Cost accounting on a per-unit or departmental basis can often mislead managers in a lean organization. Costs of recycling, cleaning, and by-product production on a unit level can look unprofitable. The corporate philosophy of lean and its benefits are lost unless you have Ford's vision, mission, and synergy. His almost endless hiring of people to recycle and find savings seemed almost paradoxical to the new generation of managers. Ford understood that lean did not mean running with little staff and workers. It focused on waste elimination, at times using his "common sense" to overrule the accountants.

After retirement, for example, much of Ford's lumber recycling programs were eliminated, based on cost accounting, as were many of Ford's lean programs. Large numbers of workers were cut, which in turn shut down many recycling programs. Ford Company lost the synergy that had produced aggregate profits. Just before Ford retired, Toyota and Ford's engineers spent months studying his methods. What they saw was not the powerful assembly methodology, but the synergy in the company of waste elimination. Toyota would take the torch of Ford's lean philosophy and create the world's most profitable auto company in Japan. Today, American industry is returning to Ford's original philosophy through Toyota's leadership.

16

A Shared View of the Education of Our Youth

"There is nothing in our educational curriculum, and seemly there is no place in it, for creative minds."
— George Washington Carver at the Chemurgical Conference in Dearborn, 1937

"We want artists in the industrial relationship."
— Henry Ford

Carver is often overlooked as a great teacher because his passion for research had priority. Carver, however, had research as the foundation of his real teaching. He was commonly asked to speak at teachers' conferences with Booker T. Washington. He reviewed many of his ideas at the Alabama State Teachers Association Conference in April of 1901. Booker T. Washington wanted Carver to teach more and do less research. He also often chastised him for being a poor manager of the farm. In 1911, Booker T. Washington recognized Carver's mastery of teaching and lecturing: "I think I ought to say to you again that everyone here recognizes that your great forte is in teaching and lecturing. There are few people anywhere who have greater ability to inspire and instruct as a teacher and as a lecturer than is true of yourself."[1] It is not surprising that Carver won over students as well as Congress when he spoke. He was a very gifted and passionate presenter, but his heart was in the laboratory.

One of the best testimonies of his teaching ability was described by then vice president of the United States, Henry A. Wallace:

> Because of his friendship with my father and perhaps his interest in children George Carver often took me with him on botany expeditions, and it was he that first introduced me to the mysteries of plant fertilization.... Though I was a small boy he gave me credit for being able to identify different species of grass. He made so much of it I am certain now that, out of the goodness of his heart, he greatly exaggerated my botanical ability. But his faith aroused my natural interest and kindled an ambition to excel in this field; the praise did me good, as the praise of a child often does. There is no doubt it is the gift of the true teacher to see the possibilities before the pupils themselves are conscious that they exist. Later on I was to have an intimate acquaintance with plants myself, because I spent a good many years breeding corn. Perhaps that is partly because this scientist, who belonged to another race, had deepened my appreciation of plants in a way I could never forget. Certainly because of his faith I became interested in things that today give me a distinct pleasure. I feel I must pay him this debt of gratitude.[2]

16. A Shared View of the Education of Our Youth

Carver the teacher often was at odds with his boss and first president of Tuskegee Institute, Booker T. Washington. Carver believed that his research and teaching were one, not separate endeavors. For Carver, if you didn't do research, you had nothing to teach. Of course, research for Carver was practical and hands on versus that of today. Carver's point was that teaching was not parroting some textbook. Still, Carver demanded the basics as well. Teaching was scholarly and practical. It was trade and skill oriented but with the flexibility of identifying the future scholar as well. Ford totally agreed with the approach at Tuskegee and was donating to the school in 1916 because of its educational approach. It was an ideal of a craftsman scholar that both Ford, Booker T., and Carver believed in.

Maybe just as important, Carver saw education as motivating future scholars through his pragmatic skills training. This was a different view in a time when the craftsman or artist was considered a different orientation than the scholar. This had been Carver's personal experience — so many had encouraged and helped him to achieve his position of professor while he excelled at crafts and art, and he wanted to do the same. Biographer Gary Kremer summarized Carver's strength as a teacher:

> Carver drew upon three unique talents as a teacher. First, he was genuinely interested in his students and made them feel they were truly important to him; second, he was excited about his subject matter, and he transmitted the excitement to his students; and third, his teaching methods combined the transmission of ideas with a practical application of those ideas, a combination that many students found irresistible. Above all else, Carver wanted his students to learn to think creatively and independently.[3]

This description could just as easily have been applied to the trade school approach of Henry Ford. They both saw education as practical and applied. Both men saw a need for discipline and routine as well. Ford arrived at the same point as Carver but from a far different path.

Ford was pulled into education as a community reformer and a concerned industrialist. Long before Ford became fully successful, Ford and his wife were helping orphaned boys get an education in Detroit. Ford's success in the production of the Model T brought a flow of immigrants from Europe and the American South. Both groups were poorly educated. Immigrants, however, were a bigger problem because of their lack of language skills. Ford told a journalist in 1914, "Foreign laborers cannot become American citizens, learn to spend more money for living and efficiently enjoy freedom and citizenship unless they can speak, read, and write English."[4] Ford also was concerned about the inability to assimilate into society, which was a particular problem for Detroit's Mexican immigrants at the time. Interestingly, this was the same lack of adult education faced by Carver with southern black farmers implementing his new methodology.

Ford addressed this weakness of language by forming the Ford English School in 1916 for his workers. The school had 163 teachers, and in 1916 alone, it enrolled over 2,700 students. The students were Europeans, Mexicans, and American blacks. Ford hired Peter Roberts, a former YMCA instructor who had run similar schools for Andrew Carnegie's immigrant steelworkers and the electrical workers of George Westinghouse. The Ford English School had a set schedule to accommodate the workers' schedules. Ford would continue throughout his life to expand adult education in his communities and factories.

Carver, like Ford, believed adult education to be fundamental to democracy and freedom. The black farmers of Alabama were isolated and illiterate much like Ford's immigrant workers in Detroit. Southern blacks lacking education were often worse off after the end of slavery. The evolution of Carver's Jesup Wagon had been created to take adult education to the farm because farmers could not read his bulletins. Carver realized that these newly freed slave families needed both adult and child education and even family education. Carver approached adult education in the same vein as Ford—simple, basic, and pragmatic. Both men had grown up on the style of the one-room schoolhouse and *McGuffey Reader*. Seeing the bigger picture of the problem, Carver and Ford added cooking and home economics training for the wives of farmers and workers.

Ford would also become a lifelong critic of public education. Both Ford and Carver loved routine and saw it as a basic premise of education and one of the problems of public education. Carver had learned his as a military cadet, and Ford had learned his in the McGuffey-based school of his youth. Ford's first involvement in the education process was in 1908 when he took in some Detroit orphans to work at a farm that he purchased as a camp. This had been a joint venture with the Protestant Orphan Home of Detroit. During the day, the boys went to a McGuffey-style school of Ford's design with an established routine and discipline. Early morning and evening were reserved for farm work. The boys followed a tight routine starting at 5:30 in the morning and ending sharply at 9 at night. Cleanliness and manners were stressed. Henry Ford showed the same interest and encouragement of students that was so important to George Washington Carver. On weekends, Clara and Henry often visited or planned trips to parks or to Ford's factories. As students reached the age of eighteen, Ford found jobs in the trades or sponsored them to attend local trade schools. Sunday services were a key part of the routine.

Ford held a routine schedule for Sunday for his Edison Institute students as well. They would attend a nondenominational service at the Martha-Mary Chapel. The service was presided over by a senior student. There would be traditional prayers and hymns and often a guest speaker. There would be a reading of the Ten Commandments as they appeared in the *Second McGuffey Reader*. Students picnicked on the lawns after the service and rode on trains or the steamboat on Greenfield's Oxbow Lake. During the week there were also daily "services" for students. One reporter noted, "Did you ever go to a church where the minister was under sixteen, and the guest speaker six? Where there was no creed? No collections?"[5] When Ford was in Dearborn, he attended daily services, sitting alone in a special section. After services, he often dropped in at the various schools.

Greenfield Village of American history was the perfect setting for Ford's McGuffey-style schools. The key element of the McGuffey approach was a foundation of American history, morals, religion, and nature study. McGuffey believed that it was American principles that were behind the success of American engineering and science. It was an education that developed heroes in all fields. It often interested students in nature as a gateway to science and engineering later, which had been the path of Ford and Carver. As the students approached seventh grade, Ford and Carver added trades and mechanical skills.

Both men saw the absolute need to mix trades, practical skills, and education for both the future craftsman or farmer and the scholar. This had also been the vision of Booker T. Washington in building the Tuskegee Institute, and Charles Schwab of the Pittsburgh steel industry with his trade high schools. Many of these students built the very classrooms they sat in. In Ford's many village industry community schools, as with his Greenfield Village system of schools, and like Tuskegee, the curriculum required a mixture of basics, trades, agriculture, and home economics. Much like the concepts of Carver, Ford required a school garden for children to work in and to study. In both Carver's and Ford's case, the schools were built by the students. Ford also required the same logic as Carver did in his 1910 bulletin, *Nature Study and Gardening for Rural Schools*. Carver, like William McGuffey before him, stated his approach, "No place can be called a school in the highest sense that has no pictures on the wall, no paint, or paint or whitewash on the buildings, either inside or outside, no trees, shrubs, vines, grasses or properly laid out walks and paths, which appeal to the child's aesthetic nature, and sets before him the most important of all secular lessons — order and system."[6] The views of rural schools of Ford, McGuffey, and Carver were in perfect unison.

Ford drew national attention to the McGuffey approach to education through his Greenfield Village and related schools. In 1934, he organized a huge service and placed a granite memorial at McGuffey's birth site in Pennsylvania. For years, Ford had agents scouring the country looking for old editions of *McGuffey Readers*. His collection of *McGuffey Readers* has over 300 books representing all the various editions and special variations. The reconstructed original McGuffey schoolhouse from Pennsylvania was a classroom in Greenfield Village since 1934. In 1938, Ford purchased, moved, and reconstructed William McGuffey's birthplace there. In addition, from the same Pennsylvania County, he dismantled and reconstructed a covered bridge in Greenfield Village. Ford even moved and enrolled a young relative of McGuffey in his Greenfield school. Later that year, Ford hosted the annual meeting of the Federation of McGuffey Societies. Ford had 800 medals minted for the visiting members. By the end of the 1930s, Ford had over 6,000 enrolled in his Greenfield and related Michigan schools. The Greenfield system functions to this day, helping needy kids get a top-notch education.

In the Greenfield Village system, grades of students were segregated into various buildings for classes, including the McGuffey School (early elementary), Ann Arbor House (kindergarten), the Scotch Settlement School (elementary), and Town Hall (late elementary). The rebuilt Armington Sims machine shop was used to teach the trades and the Webster House for home economics. The various barns were workshops for working on and repairing Ford tractors, cars, and wagons. During the late 1930s, Ford started an academic high school called the Edison Institute High School. The high school had its own building, focused on technology education, and operated its own radio station. The high school was a bit more traditional with its own library and other resources.

Ford also brought the "Tuskegee Way" to his Greenfield Village schools and apprentice training. The students maintained farm and industrial equipment. Old tractors were torn apart and worn pistons replaced. The students worked on steam engines and older equipment at Ford's museum. He often took the young boys to his farms and factories to make them familiar with the latest equipment of agriculture and industry. Ford hired the best teachers for the classroom as well. The basics were stressed. Ford's students maintained a radio station and broadcasting equipment for the schools. They built and repaired the buildings. God and faith were also central to his educational system. Services were held daily for all grades. The Bible was central to all classrooms, and Bible readings often started the day's class work. There was always time for prayer. When George Washington Carver visited in the late 1930s, a day was set aside for students to discuss God's role in choosing a career. Ford and Clara often visited the school's Sunday services and the various Sunday school activities. Ford also enjoyed nature study with the students. When Carver visited, he would lead the nature studies for the day.

Education for Ford reached the level of his passion for his museum. He commonly visited not only his Greenfield school system but community schools he supported. He always encouraged moralistic-based education that he felt was lacking in public education. He promoted the use of one-room schoolhouses and *McGuffey Readers* in his community schools in Michigan. Ford also had three schools around Sudbury, Massachusetts. He opened schools in his coal mine communities in Kentucky and Tennessee. He built an extensive school system for his South American rubber plantations. He gave financial support to one-room schools throughout southern Michigan. He built a series of schools around his Alabama plantation. He also built and supported one-room schools in his village factory communities. He opened up a convalescent grade school for his hospital. He built a nursing school at his hospital and one in Alabama. Education for skilled professionals and adult education became one of his later projects.

Ford would take the concept of practical education to the next level, linking it directly to the automotive industry and to a lifelong journey for the individual. Ford's concept of skills education evolved over many years. He had started his camp farm for poor city boys in 1908 in conjunction with local agencies. Ford had also been giving money to schools such as the Tuskegee Institute early on, but his first real Ford school was the Henry Ford Trade School in 1916, built near Highland Park. This school offered two years of general study followed by specialized training in foundry, tool and die making, carpentry, and skilled trades. This school was aimed at poor students, or students without a father who had to help their families financially. Ford made the school a profitable part of the organization as the boys sharpened, forged, and repaired thousands of tools for his factories. After World War II, Ford expanded his trade school to accept returning veterans. The Henry Ford Trade School became one of the most successful in the nation.

As these graduates moved into Ford plants, the opportunities continued. In 1923 at the Rouge River plant, Ford established the Ford Apprentice School as the next step. This school allowed workers to move into the skilled trades as well as study the overall operation at the Rouge operation. Ford broke new ground by allowing blacks to enter the skilled trades at his plants. This advanced apprentice school did require a trade school background or experience in the trades. A worker could choose one of six trades: machinist, steam engineer, toolmaker, die maker, electrician, or hydraulic technician. Furthermore, after four years of apprentice school, one could move on to an in-house journeyman program. In addition to these formal programs, Ford offered extensive free courses for all workers in subjects such as mathematics, chemistry, mechanical drawing, and introductory business courses. Ford believed education was the heart of success in business, and he expanded the concept of education to its most democratic application.

One of the more overlooked Ford educational programs was one for the disabled. Ford was a leader in this country in the training of the disabled to do regular jobs at regular wages. Ford expanded his Henry Ford Trade School to address this issue. His interest in the disabled had started in 1911 when he hired a manager to look at how to assimilate disabled men into the workforce. Following World War I, Ford had 670 legless men, 2,637 one-legged men, 715 one-armed men, and 10 blind men trained and working regular-wage jobs. During and after World War II, Ford increased these numbers and trained many to work in the higher-skilled trades. He built Cherry Hill ignition plant to fully employ handicapped employees. When he ran out of space at Henry Ford Trade School in 1944, he opened a temporary school to bring in another 120 disabled veterans. Furthermore, Ford helped promote the use of disabled workers in other industries such as his fellow vagabond's Firestone Tire and Rubber.

Ford also pioneered internal business training to maintain the Ford culture and methodology. In the 1920s, as Ford Motor Company expanded internationally, Ford developed the Ford Service School. The school was to train foreign-born employees to work in his branches abroad. Ford designed a dynamic type of training by which the student progressed from department to department, with roaming instructors checking progress. In addition, students were given some classroom instruction on corporate culture and methodology. It was a two-year course. Ford reported the following in 1926: "At present 450 students, many of whom are college graduates. They include 100 Chinese, 84 Hindus, 20 Mexicans, 20 Italians, 50 Filipinos, 12 Czechoslovakians, 25 Persians, and 25 Puerto Ricans. We have on the way Russians, 25 Turks, and a group from Afghanistan."[7] It's a model we see today in Toyota, Disney, and McDonald's.

Another educational institute that Henry Ford got involved with in the early 1920s was the Berry School near Rome, Georgia. Returning from a Florida vacation, Henry and Clara stopped at the Berry School. They had heard much of this school that combined agriculture and academics to train poor southern students. It had received a large donation of $50,000 from Andrew Carnegie and had been visited by President Theodore Roosevelt. The Berry School was a reflection of Tuskegee with 600 working students in both elementary and high school. Using their own brick kilns, the students built the school. They had their own sawmill to convert logs to lumber. There were textile mills, and girls trained in the art of making clothes. The school's shield was a powerful symbol of the approach with its four symbols — the Bible, the lamp for learning, the plow for labor, and the cabin for simplicity. It was the true embodiment of Ford's (and Carver's) view of education. Clara Ford developed a relationship with the founder, Martha Berry, which would last a lifetime.

The first visit so impressed Ford that he sent two tractors, a Model T, a rock crusher, and other needed farm machinery. Clara made a gift to build a girls' dormitory shortly afterward. The school soon became a stopover for the Fords on their winter trips to Fort Myers, Florida, and their Ways, Georgia, home. Henry had a special railroad spur built

The George Washington Carver School at Ways, Georgia (Bentley Historical Library, University of Michigan, File HS8581).

from his Ways home so he could park his railroad car, *Fair Lane*, at Berry. Over the next ten years, the visits and donations increased. Henry built one of his "Martha-Mary" chapels, a gymnasium, a dining hall, a water plant, and other Gothic school buildings. Ford had all the mules and horses eliminated with a gift of twenty tractors. He added to the school's farm acreage as well. Ford created a beautiful Midwestern-type campus in the fields of Georgia. He fell in love with the area, making more visits to both the school and his nearby Richmond Hill mansion.

In the mid–1930s, when Henry and Clara wintered at their Richmond Hill (Ways, Georgia) mansion, Ford also built a complete school system for the county. At Richmond Hill, Ford upgraded a very poor black school system, giving opportunities not available for black students in the South. His Richmond Hill school system was based on the one-room system with ancillary resources. At Richmond Hill, Ford had to invest heavily in a health system to support the students of these poor black families. The capstone of the Richmond Hill system would be the school he named after George Washington Carver.

Ford was spreading his educational approach throughout Michigan as well. He donated liberally to the smaller communities around his village industries. He rebuilt schools and paid teachers' salaries in many of these communities. In return, he demanded an allegiance to his approach. That approach was centered on the basics of the *McGuffey Readers*, and Ford favored the one-room schoolhouse as well. He also favored including home economics, nature study, agriculture, and mechanics. Field trips, gardening, and nature study were part of the curriculum. Still, Ford believed in basics first before any high tech subjects of the time. The McGuffey system was inspirational to create leaders in all fields. Henry Ford championed McGuffey and his *McGuffey Readers* until his death.

Ford continued his quest for industrial and agricultural education in England as well. He was one of the earliest pioneers of a degree in agricultural engineering. At his factory in Manchester, he established a trade school to prepare students for a degree in agricultural engineering. Ford purchased a stately mansion to house his Institute of Agricultural Engineering. The school offered training in all phases of advanced farming including courses in tractor mechanics, machinery, farming, crops, animal husbandry, bookkeeping, and tractor engineering. Ford also supplied scholarships for live-in students at the institute.

17

The Last of the Victorians

"In the engineering laboratory at Dearborn we are equipped to do almost anything that we care to do in the way of experiment, but our method is essentially the Edison method of trial and error."

— Henry Ford

Much of the criticism of Ford and Carver from science and engineering professionals resulted from their perceived lack of scientific methodology. They were part of the transition in science that took men out of the Victorian home laboratory and into the university research department. Both men cared little for pure theory but loved applied science. Carver takes more criticism than Ford because he lacks patents and the financial results of a Thomas Edison, yet as Ford believed, Carver was cut from the same cloth. The problem is that most tried to evaluate Carver in the framework of a university career scientist of the Edwardian era of the time. Carver was an agricultural engineer, or maybe more correctly, a chemurgical engineer of the Victorian mold. He did not advance science but creatively applied it. His methodology was the same as Edison's, but unlike Edison, he did not evaluate the personal gain of a project but was driven by a higher and simpler mission. Ford's image as an engineer suffered from his Victorian approach as well. He preferred models to blueprints of the day. He was more of a "tinker" than a designer. Both Ford and Carver were Victorians in an Edwardian age of theoretical science and university-driven research.

Victorian scientists and engineers were observers, modelers, collectors, experimenters, and classifiers. They loved fieldwork and tinkering. They were driven by the senses, not unseen theoretical concepts. Carver even challenged some early plant classification systems, wanting to classify plants on their relationships to other materials or economic uses. Carver often talked of economic groupings of plants. Ford reasoned similarly. Ford could spend endless hours trying to identify birds in different classifications, sometimes more interested in grouping them by what they ate or where they lived versus species attributes. Ford also had a love for the industrial experimentation of scientific management. Science was physically better represented by collections than theories for these men. Their science was explained and visual like that of Jules Verne versus the Edwardian writer H. G. Wells, who often used the unexplained theory for his stories.

Carver and Ford remained relentless experimenters as well as Renaissance men. They chased their passions and moved into any field that might interest them. They were men for all seasons in the mold of Francis Bacon, Thomas Aquinas, da Vinci, and Newton. They rarely specialized in any single field. They had many successes but like all innovators,

17. The Last of the Victorians

they also had their share of failures. Their closets were full of exotic failures such as Ford's gasoline-powered railroad car, and Carver's peanut oil treatment for polio. But part of their genius was a childlike approach to science. They loved experimenting, collecting, and categorizing. As they got older and less relevant to the mainstream scientists, they were called foolish and were viewed as childlike. Both almost loved the term "foolish" as inspiration. This was their charm with the average folk who saw them as sages. The root of much of their success was to go forward when others saw only a dead end. It is why they found kinship with Thomas Edison, who spent years trying tens of thousands of different materials for lightbulb filament, and Luther Burbank who classified and developed thousands of hybrid plants. Burbank was proud of his 3,000 experiments to develop various plant hybrids. Carver also had thousands of fungi in his mycological collection. Ford once wanted to hang a sign over his engineering lab — "for damn fool experiments."

Carver and Edison had the iconic Victorian laboratories stuffed with thousands of chemicals and materials from all over the world. Edison was famous for his thousands of bottles of chemicals. He so loved his chemicals that they were the only items Edison cherished and would not fully release to Ford's museum until after his death. Ford, of course, had financed the preservation of the Edison lab in Greenfield Village. Similarly, Carver's last project was the preservation of his lab with its endless collections of plants, rocks, and animal parts. Carver spent most of the last two years of his life in this preservation project. He was afraid that his beloved stuff would become a heterogeneous collection with no significance. This diversity, or even lack of organization as it might appear to the outsider, was characteristic of Victorian scientists. It arose from the experimental versus the theoretical approach to science of the Victorians.

The Carver museum at Tuskegee would preserve the famous lab. The institute as well as Carver himself funded it. Carver would donate his life savings of $60,000 to the museum foundation. It would also be a personal museum, and much like Ford's, eclectic in nature. This amazing gift came from a professor who had made as little as $1,000 a year over his forty years at Tuskegee. Carver lived his lean research in a frugal existence.

Carver took on this last project with a great deal of enthusiasm, clearly inspired by Ford's museum. In fact, like Ford, Carver looked at the museum as an educational project for the future. Like Ford's museum, the Carver museum involved many personal accomplishments and mementos such as the skeleton of the first oxen, Betsy, used at the experimental station, and two-foot jars of the first fruits of the station. Besides thousands of rocks, plants, fibers, soils, clays, insects, and products, there were personal items such as his paintings, wallpaper designs, knitting, and other handcrafts. Like the Ford museum, Carver's museum had strong themes mixed with personal passions. They were pure Victorian.

It would be a mistake to write Carver off as a true scientist in any era. Carver could have been a university scientist of note. His many letters over the years to scientific friends show a deep understanding of mycology and its scientific terms. Carver had been asked by the government to participate in several major field studies in mycology, and samples came from around the world for Carver to classify. While chemistry and geology had advanced beyond the classification stages, mycology remained in the Victorian world.

But even as the Victorian world ended in the 1910s, great observers, collectors, classifiers, and experimenters were still needed in science. Victorians always dreamed big. Carver was not the new type of university scientist adding small bricks of research to a great wall. These university scientists are necessary to slowly build the body of knowledge, but not all scientists are suited for this type of career. Dreamers were and are still needed to inspire scientists. Carver was a big-picture dreamer like Ford, Edison, Burbank, and other Victorians. Victorian scientist always saw their roles in the bigger picture as part of a mission or Providence.

The Victorian methodology of Carver, Edison, Burbank, and Ford was an experimental-based observation. Edison's methodology, in reality, was no more "scientific" than that of Carver. In Edison's early career, he too lacked creditability and acceptance from the scientific community. Edison, however, focused on financial payback from the start of a project. He had learned the hard way with his first invention of the automated vote counter that invention without payback was a lost effort. Edison researched his projects thoroughly before committing his time and resources. He was late entering the competition of the incandescent light because of his economic evaluation of the project. Edison was also a great closer and self-promoter of his successes. Edison was a true showman. Still, his deductive and experimental methods were the same as Carver's, with the exception that Edison kept extensive laboratory records. Of course, Edison was better funded in the end and had an army of engineers and scientists working for him. Furthermore, Carver lacked the marketing and self-promotional skills of an Edison. Carver's lack of commercial success was his greatest disappointment, not because of the income, but because in a Victorian world, science was measured by its commercial success.

Still, what made Victorians like Carver, Edison, and Ford truly successful was their creativity more than their pure science. Carver correctly called himself a trailblazer versus a scientist. Ford saw himself in the same light. The new breed of university scientists and engineers were trained to advance science in small, steady steps, putting bricks in a great wall. Men like Carver and Ford dreamed of the wall itself. They were creative architects versus bricklayers. University engineers were often poor mechanics, but the modern world needs both. Victorian scientists looked to breakthroughs versus slow progress. Creators like Carver and Ford were a rare breed. The early Victorian engineer and scientist had this type of big-picture creativity often lacking in today's best scientists. More recently, we have seen this in men like Steve Jobs. Jobs specialized in creativity versus engineering. His vision was greater than the science or the engineering of computers. He often asked more of his engineers than they could envision. Jobs looked to the breakthrough and often was frustrated with mere progress. He was also a "Victorian" in his vision, which was not limited by the formulas, blueprints, and devices his creativity improved on. In reality as seen by Steve Jobs, the "Victorian" nature of Carver and Ford is ageless.

Their Victorian creativity stills leads to breakthroughs today. This is why the biographies of men like Carver and Ford are so necessary today. It is their creativity that deserves further study more than their achievements or personalities. Scientific advances will always have a degree of trial and error, experimentation, and sometimes even serendipity. Ford's industrial experimentation strengthened the scientific management movement.

17. The Last of the Victorians

Many have called Ford a social engineer, which is a good description of the Victorian approach to management. Victorians had seen the exponential growth of engineering by applying science and hoped to see the same in management by applying scientific principles. While the hard sciences show a reliance on theory over experimentation, management science still requires a great deal of experimentation to test out ideas. Creativity in the arts still requires experimentation. Today what was science in Victorian times has split into basic science and engineering. Engineers today experiment with basic scientific principles to make things. In today's terms, Carver was a materials engineer versus a scientist. Henry Ford was an industrial engineer and scientific manager.

Even in the science of management, which Victorians pioneered, they maintained earlier approaches of simple cost analysis and process experiments. Victorian industrialists such as Ford, Carnegie, and Rockefeller had created cost accounting and used it to drive production costs down. Ford had carefully watched the cost streams that went into manufacturing as a means of reducing them. Victorians, however, did not like the "theoretical" distribution of overhead onto products costs. They also commonly ignored opportunity costs. They considered a product profitable if, in today's terms, it contributed to margins. Based on this approach, Ford's salvage operations, small product lines, and such appeared more profitable. Not that overhead costs were ignored, but they were viewed as an aggregate factor in the overall profitability of the company. That is why Ford often was very satisfied with his village factories where costs were well defined and measurable. Like modern scientists and engineers, modern accountants tend to look at Victorian accounting methods as rather backward.

In science and engineering, these Victorians proved very creative and successful. In fact, the Victorian era from roughly 1830 to 1930 is unmatched in history for advances in science and technology. The question becomes, are Victorian methods irrelevant in today's high-tech world? In fact, one of the detailed statistical studies of human accomplishments in science and technology suggests a decline in the rate of advance with the end of the Victorians in 1950.[1] The reasons for this are not clear and may have nothing to do with Victorian methodology. One theory is that every branch of human endeavor reaches a golden era followed by a period of slower advance. Victorians were characterized by some unique approaches. They were goal driven and achievement oriented. Victorians also focused on breakthrough goals. Edison set a goal for his lab to produce a major invention every six months and a minor one every six weeks. They were capitalists hoping to gain financial returns from their efforts. They loved competition and found pleasure in being first or the best or producing the biggest or largest of something. The Olympics of their day were the great World's Fairs of technology. Some set their own goals and records to break; for Carver, there was a thrill in expanding his lists of uses for peanuts, sweet potatoes, and other plants. He was included in major government classifications of fungi because of his skills in identification and his ever-expanding lists. Classification skills became important in Carver's experimental methods.

While Carver believed in the same methodology as Ford and Edison, he lacked the financial resources to fully develop them into commercial successes. Carver was not as interested in personal financial gain as much as to see the black farmers benefit from his products. Carver's goal was to get his methods, recipes, and new crops adopted by southern black farmers, and that required marketing. Carver, like Ford and Edison, had a down-home understanding of product development and marketing. Carver talked of three stages of market development for new products:

Stage 1: public hostility and rejection of something new. He described it as people sticking to their old and familiar ways. Certainly, Carver knew this stage well with the farmers of the Cotton Belt. "The new product must meet successfully all hostile tests. It must prove its superiority."[2] Carver learned this in decades of opposition to change from a cotton-based economy. Ford also faced this phase; and in Ford's case, he used auto races to prove out the superiority of horseless carriages. Ford, however, entered the horseless carriage market late, letting others fully work through this phase.

Stage 2: stage of total apathy. For Carver, this stage was often decades long before some of his early ideas took hold. His sweet potato flour lingered in this stage for decades, after an initial trial in World War I, to become an important help in World War II. Many of Carver's lab developments, such as cottonseed, soybean, and peanut oils, often had to await better processing equipment for full commercialization. It was at this stage that education and marketing were critical tools. For many decades, southern farmers held on to cotton as their main crop; it took a generation of reeducation by Carver to change things.

Stage 3: new uses, improved product attributes, and related discoveries (financial rewards). While Carver had learned to evaluate projects for possible uses before entering the quest, Carver's interest was in the first and second stages, not the realization of financial rewards. Carver worked from a perceived need for something, not what he could personally gain. Edison loved to jump in at this stage where the money for a new product could be realized. Edison often took six months to a year to explore the financial potential before doing any research. Edison was the last to enter the quest for a usable incandescent light-bulb, but he was the first to develop a commercial one. For his part, Carver based these stages on the history of civilization, which he defined in terms of the advance of materials. He argued that the three stages of history were the finding, adapting, and finally chemically improving the earth's raw materials. Ford made his mark with affordable mass-produced cars, entering the field late but correctly identifying the needs, wants, and product problems of the consumers to produce a car for the masses.

While one can make a distinction between Victorian experimental and theoretical science, the clearer distinction is in the end goal. Victorian science had always been focused on the end goal and commercial benefit. Theoretical science focuses on the advance of theory. Even Carver was focused on the economic benefit of the farmers. Henry Bessemer had set out to invent a steelmaking process, as Edison had with the electric light, Martin Hall with aluminum, and Alexander Graham Bell with the telephone. There had been little pure theoretical work in science prior to 1920 because, while more valuable in the long run, it lacked short-run paybacks. The lack of immediate commercial gain and the need for specialized researchers forced science into the universities. Money prizes such as those of Napoleon III had often inspired focused Victorian science. World's Fairs had been the visual and commercial demonstration of the success of Victorian science. Henry

Ford never changed much from his Victorian approach since engineering was less affected by the transition in the sciences. However, as a scientist, Carver was caught in the transition of science to university research teams.

Chemurgy was an example of science in the transition from Victorian to theoretical and the individual to the university. Chemurgy was a mix of science and applied science. But even before the chemurgical movement, there was the "creative chemistry" movement. The "creative chemistry" movement was transitional in nature. Edwin E. Slossan coined "creative chemistry" in the 1919 book *Creative Chemistry*. Creative chemistry embraced both experimental and theoretical chemistry. Carver preferred the title "creative chemist" to that of stovetop chemist. While Carver and others called him a "stovetop" or "cookstove" chemist, Carver learned to see it as belittlement. Biographers have often spent a great deal of time with Carver's transition from "cookstove chemist" to a "creative chemist." The transition or transformation, however, was one of perspective rather than some personal objective of Carver's. Carver, like Ford, is striking in his lifelong consistent approach to his work. Clearly, Carver wanted the recognition that Edison had in his prime, but Carver didn't change to gain that recognition. He tried to argue for his methodology, never backing down from his asking for divine guidance in the laboratory or his simple experimental approach. Carver's mixing of religion with science destroyed his creditability with many university scientists.

Another part of Victorian science was a respect for or at least recognition of Providence, a supreme power, or a creator. Post-Victorian scientists believed in the elimination of religion and all talk of God in the context of science. The post–Victorians at the time were locked in a battle over evolution. Scientists became sensitive to any mention of God in the same framework as science. Carver, of course, had brought much criticism on himself by mixing the two. His mention of God's inspiration in the laboratory brought an attack by the *New York Times* on Carver's creditability after a talk in New York. Many pointed to Victorian scientists such as Edison as being silent on the issue as a proposed norm for scientists with religious beliefs. Men like Edison and Ford, however, had a non-denominational spirituality that was at best barely acceptable at universities and was more typical of the later Victorian age. When pressed, both Ford and Edison acknowledged the role of a God in their success. Ford demanded prayer at all his educational programs. Edison was interviewed about the role of a supreme power in his discoveries. He affirmed that it was God that was behind his success.[3] Still, Carver's enthusiastic talk of God in his experimentation seemed dangerous to these modern theorists. Theory had to be free of any religious link in their opinion. Carver's faith seemed to rule him out as a theoretical scientist, but there were other reasons to rule out Carver as a scientist in general. He based his work on experimental results, with theory only as a way to begin the search.

University scientists believed that theory was the beginning and end of science. The creative scientist was an applied scientist or an engineer. Synthetic fertilizers of the 1920s were often cited as the combination of theory and experimental work to produce commercial value by creative chemistry. This was creative chemistry at its best, applying theory to design experiments. Today most universities make the distinction in their science versus engineering curriculums, in that science is theoretical and engineering is applied science. Carver and other creative chemists and later chemurgists all used the basic theoretical concepts of chemistry in their initial work. Like today's engineers, these creative scientists and chemurgists were trained in science, but they applied these basic principles versus improving the theories. By the 1950s, the theoretical science would become chemistry and the applied science that of the chemical engineer, environmental engineer, or materials engineer. In this respect, Carver was really one of the early chemical and materials engineers. Trying to rate Carver in terms of a university scientist leaves him lacking in the advancement of theory, but it is an artificial weakness of no real consequence in his contribution to society.

One Victorian scientific attribute that Carver appeared to ignore was that of lab journals and notebooks. Written notes were critical to observation-based Victorian science. Carver's lack of notebooks and formulas had often concerned visiting scientists looking to commercialize his products. The reasons for this paperless methodology are not clear, but it was his way. It may have been part of his thrifty nature. As a student, he often lacked money for paper and pencil and made his own paints from clay. He had even developed from plants de-inking chemicals to renew paper. Even his beloved paintings were painted and drawn on cornstalk canvas. He learned to remember his methods, summarizing them after years of experimentation in his bulletins and newsletters. This was similar to Henry Ford who preferred models to engineering drawings. It was, however, much different than Thomas Edison who loved to fill shelves of notebooks. Many believe he was secretive about his discoveries, but it appears more an operating style versus secrecy. Carver had proved to be a fairly prolific letter writer and many commercial and technical discussions are included in these. In particular, his lifelong letter writing to mycologist Doctor Pammel at Iowa State is often scientific in nature and enclosed exchanges of scientific specimens. He often supplied specific recipes to people who wrote him or asked him to identify a plant or analyze a soil specimen. There was a period when he was trying to commercialize paints and other products where he preferred some secrecy. His lack of lab book records often inhibited his ability to obtain patent rights. Still, had Edison run Carver's lab, there would have been a long list of patents.

In his last years, Carver received many of the honors that had eluded him for his scientific work. He received the Roosevelt Medal for distinguished service in the field of science in 1939. In 1941, the University of Rochester followed Simpson College in giving him an honorary doctor of science. Many awards in science followed in 1942. Just as numerous were his humanitarian awards, which poured in as the world feared his loss. More than anyone, Henry Ford made sure that Carver received the recognition he had earned. Ford made Carver's home a shrine in Greenfield Village and named schools after him. In the village, Carver took his place with Ford, Edison, Burbank, Heinz, the Wright Brothers, and others as a great American innovator. Thanks to Ford, the spirit and continuing legacy of these innovators live today for future generations.

18

Continuing Legacy

"Well, someday I will have to leave this world. And when that day comes, I want to feel that I have an excuse for having lived in it."
— George Washington Carver

Both Carver and Ford died a quiet death. Interestingly, a sip of cow's milk was the final food for both of them and one they had tried to replace. Both men died at their homes. For Carver, it was January 5, 1942, at Tuskegee; Ford died on April 7, 1947, at his Dearborn home. Carver, of course, died without family but with friends at Tuskegee. Ford died the same day he visited his village industry at Flat Rock. Condolences flooded in from around the world for both men. President Franklin Roosevelt and Vice President Wallace, both of whom were Carver's friends, wrote eulogies for him. Then Senator Harry Truman introduced a bill to build a national monument. Carver was laid to rest at Tuskegee Institute. Ford's death was marked by the stopping of assembly lines around the world. He lay in state at his beloved Greenfield Village and was buried at Addison Ford Cemetery in Dearborn. These deaths marked the true end of the Victorian age of science and technology in the United States.

Direct comparisons of the lives of Carver and Ford are difficult to make because of Carver's starting point and his having to deal with racism throughout his life. An editorial in the British magazine *Efficiency* put it best: "If I were asked what living man had the worst start, and the best finish, I would say Dr. Carver. It is a great loss to us that we have no one like him in England." Henry Ford would have agreed. The legacy that both men leave is one of creativity. That creativity lives today in lean and green manufacturing as well as industrial agriculture. Both men founded the chemurgical movement which has morphed into many green and environmental movements. Ford and Carver pioneered an industry-friendly green approach that could unite the many fractions of today. Both men pioneered lean manufacturing as well. Industry now spends millions to apply the lean manufacturing principles of both men.

Chemurgy is returning with vengeance, but don't look for the term to return. Today it is called biotechnology, ecology, green, or bioengineering. The *Economist* hailed chemurgy's return in 2008: "The chief reason for such optimism is that industrial biotechnology is better and cheaper than back in the heyday of chemurgy. Dow Chemical has even come up with a material made from soybean oil that it plans to sell to carmakers to replace oil-based foam. Ford and his friend Carver would be proud."[1] Interestingly, we have gone from Spanish moss to stuff Model T seats to petroleum-based foam, and now to chemurgical soybean-based foam. It is even more amazing that the old chemurgy

Carver's meeting with President Franklin D. Roosevelt in 1936 (Alabama Department of Archives and History, Montgomery, Photograph Q4849).

promoters of Ford Motor Company and Dow Chemical are on the forefront of the return of chemurgy.

Many of the old research papers of Ford and Carver are once again being looked over. Ethanol is being used at the 10 percent level in our gasoline, and there is an effort to increase that percentage. We are approaching Ford's ideal mix of 30 to 40 percent ethanol. Brazil is fueling its cars with near pure ethanol. Carver's alma mater, Iowa State, is once again leading the nation in ethanol research as it did in the 1930s with the support of the Farm Chemurgic Council. Ford and Carver both advocated using fewer sugar-based sources and more cellulosic sources for alcohol production such as cornstalks and grasses. Iowa State today is looking at switchgrass as a future source for commercial ethanol. Cellulosic alcohol had been pioneered by Henry Ford and George Washington Carver. Ethanol factories are adopting the zero-waste policies of Ford and Carver with all parts of the corn crop being utilized. Soybeans are once again being used to produce biodiesel fuel as are other vegetable oils. Finally in 1986, Henry Ford's concern for the use of carcinogenic tetraethyl lead was realized and banned. But had we adopted ethanol mix, the atmosphere would have been saved sixty years of lead pollution.

Henry Ford's environmental ideas are finding new support. Ford's belief in clean and green energy is once again on the front pages, and his company has a leadership posi-

18. Continuing Legacy

tion again in the green movement. Ford Motor has been the world leader in the implementation of the ISO 14000 environmental standard. Today we see clean fuels of Ford's early work finding applications. Lightweight plastic car parts such as Ford pioneered in his soybean car are saving fuel. Ford factories today have greenhouses on the roofs. Ford has leadership in the use of recycled fabrics in its Focus and Escape models. Ford's belief in electric cars is again resurfacing.

Similarly, Carver's ideas are having a rebirth. Carver's use of recycled motor oil is standard practice today. Carver's recycling ideas have become mainstream in homes and farms across the nation. Carver's composting has led to a rebirth today in organic gardening. America is once again looking to energy independence that Ford and Carver saw in American agriculture. Ford's belief that a shortage of raw materials was the motivation for wars has proved prophetic as seen in the Middle East today. Ford's lean approach to manufacturing is being practiced the world over. Toyota has given us a second revolution in manufacturing based on Henry Ford's original lean principles.

Lean and green principles are bringing back the harmony of manufacturing and the environment that both Ford and Carver espoused. The greatest legacy of Henry Ford's chemurgical efforts is that Ford Motor today has revived Henry's work in conservation, alternative fuels, recycling, and green initiatives. For over a decade, Ford Motor led industry in demanding the international environmental standard ISO 14000 from its suppliers. Ford plants often have gardens and solar panels on the roof. Its River Rouge plant has a ten-acre living roof. Ford Motor has made huge advances in using plastic and aluminum to improve fuel efficiency just as Henry had envisioned.

Conservation and use of scrap wood is again returning. Much like Ford's obsession with saving wood, a southern company has been formed to make charcoal from bush and discarded tree clearings based on Ford's early work on Michigan farms. Using small portable kilns, this company can go on location making various grades of charcoal. Similar to Ford's work in clearing Michigan farm land, which led to the salable product of charcoal, this company is saving precious resources and helping to cut charcoal imports to the United States and the western world from South American tropical forests, which are being used up at 10,000 square kilometers a year for charcoal.[2]

Plastics from nature are once again being seen as a renewable industrial product. In 2010, Ford Motor resurrected a straw-reinforced plastic similar to the Fordite of the 1920s. This composite is being used in the Ford Flex today to reduce weight and save fuel. Soybean plastic foams are once again being used in most Ford cars today. Ford Motor is using over 30 million pounds of recycled plastic in its cars. The company has once again launched a huge research effort in plant-based materials and recycled materials. Ford Motor is even talking about a "bio-car" in similar terms used by Henry back in the 1930s. Plastic body and framed cars are again making the news.

Ford and Carver's soybean resin composite plastic is once again being used. Soybean resins composite is finally making the impact that Henry Ford foresaw. In 2002, John Deere began making panels for combines out of soybean- and corn-fiber polymers. These soybean composites are as strong as steel and weigh 25 percent less. Of course, the use of these panels has reduced fuel consumption significantly for farmers.

The chemurgical work of both Carver and Ford in fibers and composite materials is continuing today and may well be one of their greatest legacies. Fiber such as hemp, okra, cotton, and glass in vegetable plastic resins can exceed the strength of metals like steel at a fraction of the weight. Industrial research in hemp is being permitted in the Midwest again. Ford and Mercedes-Benz are once again using hemp in plastic auto parts. General Motors also purchases some hemp parts. As Carver had suggested, industrial hemp is today being used in insulation for building. Adidas is using hemp fiber in shoes and Calvin Kevin is using it in textiles. Hemp is being studied as a nonwood additive to papermaking. While more expensive than fiberglass, hemp is a great insulating material for eco-conscious consumers. Fibers such as hemp and flax are once again getting a second look because of the strength and durability noted by Carver and Ford. Hemp paper again is a growing material that can save millions of acres of forest. Ford had correctly predicted that deforestation could be reduced by 50 percent by switching to hemp paper. Hemp oil is once again being considered for biofuels as Henry Ford had suggested in the 1930s.

Okra fiber was another product considered because Carver had made extremely wear-resistant rugs of it. He was the first to recognize the strength of okra fiber. Only recently have engineers found that the tensile strength of okra fiber is on the level of glass and are using it in polyester composite materials that approach the strength of fiberglass. Ford and Carver both advanced fiber-backed resins as the future of materials. Carver also advanced the manufacture of bonded-type leather products from these plant fibers, which once again are finding applications. This was all part of the science of composite materials in which Ford and Carver were pioneers.

Composite materials are saving the fuel that Ford had promised with his soybean cars. These composite materials are light and strong and are finding new applications in transportation. Composite-type natural materials that Carver and Ford both worked on are returning in alternative disposal materials such as palm leaf plates and bowls made from leaves and resins to replace Styrofoam. Ford's and Carver's pressed resin products such as fiber plastics, soybean plastics, and casein plastics are now commonplace. Ford and Carver were clearly earlier pioneers in the fiber-strengthened materials of today. Glues similar to Carver's sweet potato glue are the mainstay of the plywood and pressed board industry.

Ford and Carver's work in the 1940s on soybean research for food products proved visionary. They both had worked decades on milk substitutes from soybeans and peanuts. Much of this research would lay the groundwork for today's substitute dairy products such as coffee creamers and whipped dessert toppings. Today, soybeans are a mainstream American crop. Current research is in the potential use of soybeans in biofuel production for diesel engines, which Ford had called for first. Today soybean ink pioneered by Carver is being used in most newspaper print because it prints more per page with better color. Soybean crayons are being made because they hold colors better than paraffin wax. When Ford and Carver were born, there was no acreage of soybeans in the United States; at their deaths, farms were producing 200 million bushels a year. Similarly, peanuts had gone from a novelty to the South's second-biggest crop after cotton, with more than 5 million acres in production. At Carver's death, the peanut industry was a $60 million a

18. Continuing Legacy

year industry. In general, Ford and Carver had fundamentally changed what American agriculture grew and produced. Today, Ford's Brazilian rubber town of Fordlandia is a ghost town, but the area is having an agricultural boom, ironically thanks to soybeans being introduced. Brazil ranks second after the United States in soybean exports.

Ford's dream of running cars and tractors off soybean oil is being achieved. Diesel biofuels are making a huge dent in fuel usage. City buses and taxicabs are commonly running off vegetable oil just as Carver and Ford had once predicted. Five to 10 percent soybean oil/diesel mix is becoming common throughout the world. Chemurgic work is progressing in rubberlike materials again. Research is even being done in the use of dandelions and other plants to make rubber. The old chemurgic debate of "can we grow enough" has surfaced again. The limitation of crop acreage shorted Ford's full implementation of the soybean car and fuels.

Soybean plastic car bodies also achieved another dream of Ford's to build a rustproof car. He had experimented with aluminum and had hoped that the dream might be achieved in the 1920s with cheap electricity needed for aluminum production. His failure with his Muscle Shoals power plant ended the dream. In 1935, he built six stainless steel cars that were also rustproof but over ten times the cost of the best steel. But molded soybean bodies offered a major saving and the elimination of huge, costly stamping machines.

Maybe more amazing is that the McGuffey-style education promoted by Henry Ford and Carver is getting another look. *McGuffey Readers* dominated American education from the 1840s into the 1900s, including the educations of Henry Ford, George Washington Carver, Harvey Firestone, and Thomas Edison. Today there is a revival of their use among homeschoolers. Sales of *McGuffey Readers* in 2011 were over a quarter million copies. The simple basic approach is once again showing progress. McGuffey jingles such as "Twinkle, Twinkle Little Star," and "Where there's a will, there's a way" remain in the vernacular even today. *McGuffey Readers* were best sellers only surpassed in America by the Bible and Webster's dictionary. In 2008, the *McGuffey Eclectic Reader* was ranked with Thomas Paine's *Common Sense* and Alexander Hamilton's *The Federalist* as "books that changed the course of U.S. history." Henry Ford's collection of hundreds of *McGuffey Readers* remains the best reference source.

The public legacy of Carver continues to approach that of Henry Ford. He had a national monument dedicated to him in Diamond, Missouri. He appeared on two commemorative stamps. He had a nuclear-powered submarine named after him as well as dozens of schools. He is in the Hall of Fame for Great Americans and the National Inventors Hall of Fame. Carver's role in changing America's view of racism is undeniable, but it was a product of his scientific achievements.

What is unfortunate is the type of postracial perspective of Carver's work seen today. Carver's work is called exaggerated or overblown by these new historians. Even worse, they see Carver's achievements as part of America's transition from Jim Crow to black equality, his scientific work being, in part, the fictional product of social reformers. Carver becomes a symbol not a scientist. These views are not correct. Carver's impact on race relations was in his work. His scientific achievements changed the view of blacks in the nation. He never wanted to be the symbol or change agent of a Booker T. Washington or a Doctor King. By being a good scientist, he changed race relations. He was not in leadership or part of mainstream black movements such as the NAACP; in fact, he was often at odds with their approach. He approached race as he did his Christianity, with a calm inner belief in the future. His legacy is correctly discussed in the halls of science versus sociology, race relations, or politics.

Maybe the best legacy of both men is their example of what American capitalism can be. Capitalism brings democracy to the marketplace. Carver saw a freedom in the jobs that manufacturing could bring to the South. Ford and Carver saw capitalism as of and for the people. For them, American capitalism was built on manufacturing and agriculture, not banking. In fact, they both saw bankers as roadblocks to bringing capitalism to the people. Both men saw capitalism as national and local, not on an international level. Carver had gained notoriety in his congressional speech for tariffs to protect American agriculture and manufacturing. Both men were believers in the chemurgic "Declaration of Dependence on American Resources" signed in 1937. They also believed that protected national trade would lead to peace because it would force economic independence. For both men, the root of war was dependence on international resources.

Ford and Carver showed the heart of American capitalism in giving back and helping. This, of course, was one of the key lessons of the *McGuffey Readers*. Ford built schools and hospitals in the most needed areas. Carver gave endless hours to youth groups, farm organizations, and government agencies. They believed, like the first Scotch-Irish frontiersmen, that education was needed for both the individual and the nation to advance, and that education must be available to all for democracy and capitalism to flourish.

Timeline

1863	Henry Ford born in Dearborn, Michigan
1865	George Washington Carver born in Diamond Grove, Missouri
1871	Ford starts at a Scotch Settlement School at age seven
1877	Carver starts school at Neosho, Missouri, at age twelve
1878	Carver moves to a better school in Fort Scott, Kansas
1879	Carver moves in with the Seymours of Olathe
	Ford starts his apprenticeship at Flower Brothers, Detroit
1880	Carver and the Seymours move to Minneapolis, Kansas
	Ford starts as a machinist at Detroit Dry Dock
	Carver starts a laundry business of his own but remains in school
1882	Ford returns to Dearborn to help farmers repair and operate Westinghouse steam engines
1883	Carver's brother Jim dies from smallpox
1884	Carver gets typing clerk job at Kansas City's Union Depot and buys farmland
1884	Carver moves to Kansas City and starts as a clerk at Union Depot
1885	Carver is accepted at Highland College only to be turned down on arrival
1886	Carver purchases a 160-acre homestead
	Ford becomes a troubleshooter for Westinghouse Company's farm equipment
1888	Carver leaves homestead and moves to Winterset, Iowa
	Ford marries Clara Bryant and settles in Dearborn
1890	Carver enters Simpson College
1891	Ford starts work at Edison Illuminating Company
1893	Ford goes to Chicago World's Fair
	Carver's painting wins honorable mention at Chicago World's Fair
1896	Ford's first car — the Quadricycle
	Carver becomes professor at Tuskegee Normal and Industrial Institute
1906	Carver introduces Jesup Wagon
	Ford introduces Model N
1908	Model T introduced
1913	Ford's Highland Park assembly line in full operation
1914	Ford announces the five-dollar-a-day wage
1916	Carver invited to join British Royal Society of Arts
1920	Ford's River Rouge plant opens
1921	Carver speaks to Congress in support of the tariff

	Ford proposes revolutionary Tennessee Valley project — Muscle Shoals
	Northville, Michigan, first of Ford's village industries
1923	Ford buys rubber plantation "Fordlandia" in South America
1927	Model A replaces the Model T
1929	Edison Institute and Greenfield Village open
1931	Thomas Edison dies
1933	Ford introduces soybean plastics at World's Fair
1935	Ford sponsors first national chemurgical conference at Dearborn
1937	Carver and Ford meet at the Dearborn Chemurgical Conference
1941	Carver Museum is opened and Ford comes to opening
1942	Ford opens the George Washington Carver Cabin in Greenfield Village
1943	Carver dies at seventy-nine
1947	Ford dies at eighty-three

Chapter Notes

Introduction

1. John Perry, *George Washington Carver* (Nashville: Thomas Nelson, 2011), p. 146.
2. Barry Mackintosh, "George Washington Carver: The Making of a Myth," *The Journal of Southern History*, Vol. 42, No. 4, November 1976.
3. Gary R. Kremer, *George Washington Carver in His Own Words*, (Columbia: University of Missouri Press, 1987), p. 161.

Chapter 1

1. Carroll W. Pursell, "The Farm Chemurgic Movement Council and the United States Department of Agriculture," *History of Science and Society*, Vol. 60, No. 3, Autumn 1969, p. 309.
2. Proceedings of the Dearborn Conference on Agriculture, Industry, and Science, Dearborn, Michigan, May 7 and 8, 1935, p. 11.
3. Carroll W. Pursell, "The Farm Chemurgic Movement Council and the United States Department of Agriculture," *History of Science and Society*, Vol. 60, No. 3, Autumn 1969, p. 309.
4. Proceedings of the Dearborn Conference on Agriculture, Industry, and Science. Dearborn, Michigan, May 7 and 8, 1935, p.183.
5. Reynold Millard Wik, "Henry Ford's Science and Technology for Rural America," *Technology and Culture*, Vol. 3, No. 3, Summer 1962.
6. The amount has been quoted between $25,000 and $100,000.

Chapter 2

1. Linda O. McMurray, *George Washington Carver: Scientist & Symbol* (New York: Oxford University Press, 1981), p. 18.
2. Henry Ford, *My Life and Work* (New York: Doubleday, Page, 1922), p. 15.
3. S. J. Woolf, *New York Times Magazine,* January 12, 1936.
4. Racham Holt, *George Washington Carver: An American Biography*, (New York: Doubleday, Doran, 1943), p. 216.
5. Harvey Minnich, *William Holmes McGuffey and His Readers* (New York: American Book Company, 1936), p. 7.
6. Greg Grandin, *Fordlandia: The Rise and Fall of Henry Ford's Forgotten Jungle City* (New York: Picador, 2009), p. 257.
7. Douglas Brinkley, *Wheels for the World: Henry Ford, His Company, and a Century of Progress* (New York: Viking, 2003), pp. 9–10.
8. Quentin R. Skrabec, Jr., *William McGuffey: Mentor to American Industry* (New York: Algora, 2009), pp. 217–223.
9. Steven Watts, *The People's Tycoon: Henry Ford and the American Century* (New York: Alfred A. Knopf, 2005), p. 25.

Chapter 3

1. Paul Wellman, "Friends of Old Days in Kansas Saw Budding Genius of Negro Scientist," *Kansas City Star*, September 9, 1942.
2. James Wilson under Presidents McKinley, Roosevelt, and Taft, Henry Cantell under Harding and Coolidge, and Henry A. Wallace under Franklin D. Roosevelt.
3. Steven Watts, *The People's Tycoon: Henry Ford and the American Century* (New York: Alfred A. Knopf, 2005), p. 27.
4. Sidney Olson, *Young Henry Ford: A Pictorial of the Firs Forty Years* (Detroit: Wayne States University Press, 1997), p. 54.
5. Ford Bryan, *Clara: Mrs. Henry Ford.* (Dearborn: Ford Books, 2001), p. 34.

Chapter 4

1. Racham Holt, *George Washington Carver: An American Biography* (New York: Doubleday, Doran, 1943), p. 88.
2. John Perry, *George Washington Carver* (Nashville: Thomas Nelson, 2011), p. 23.
3. Linda O. McMurray, *George Washington*

Carver: Scientist & Symbol (New York: Oxford University Press, 1981), p. 41.
   4. *Detroit Free Press,* March 7, 1896.

Chapter 5

   1. Letter from George Washington Carver to Booker T. Washington, April 17, 1896, Box 4, George Washington Carver Papers, Tuskegee Institute Archives.
   2. "$600,000 for Tuskegee," *New York Times,* April 24, 1903.
   3. Mark D. Hersey, *My Work Is That of Conservation: An Environmental Biography of George Washington Carver* (Athens: University of Georgia Press, 2011), p. 117.
   4. Letter from George Washington Carver to Finance Committee, November 11, 1896, Box 4, George Washington Carver Papers, Tuskegee Institute Archives.
   5. Linda O. McMurray, *George Washington Carver: Scientist & Symbol* (New York: Oxford University Press, 1981), p. 53.
   6. John Perry, *George Washington Carver* (Nashville: Thomas Nelson, 2011), p. 41.
   7. Ibid., p. 30
   8. Glenn Clark, *The Man Who Talks with the Flowers: The Intimate Life Story of Dr. George Washington Carver,* (Blacksburg: Wilder, 2011), p. 34.
   9. Linda Hines, "George W. Carver and the Tuskegee Agricultural Experiment Station," *Agricultural History,* Vol. 53, No. 1, p. 81.
   10. Conversation with Professor G. W. Carver reported by L. H. Pammel, February 11, 1924, Iowa State Archives, George Washington Carver Collection.
   11. Racham Holt, *George Washington Carver: An American Biography* (New York: Doubleday, Doran, 1943), p. 213.
   12. Gary R. Kremer, *George Washington Carver in His Own Words,* (Columbia: University of Missouri Press, 1987), p. 6.
   13. Mark D. Hersey, *My Work Is That of Conservation: An Environmental Biography of George Washington Carver* (Athens: University of Georgia Press, 2011), p. 126.
   14. *Collier's Weekly,* April 29, 1905.
   15. Michael Lacey, *Ford: The Men and the Machine* (Boston: Little, Brown, 1986), pp. 74–75.

Chapter 6

   1. Quentin R. Skrabec, *William McGuffey: Mentor to American Industry* (New York: Algora, 2009), p. 199.
   2. B. D. Mayberry, "The Tuskegee Movable School: A Unique Contribution to National and International Agriculture and Rural Development," *Agricultural History,* Vol. 65, No. 2, Spring 1991, pp. 85–104.
   3. Allen Jones, "The South's First Black Farm Agents," *Agricultural History,* Vol. 50, No. 4, October 1976, p. 641.
   4. Ford R. Bryan, *Clara: Mrs. Henry Ford* (Dearborn: Ford Books, 2001), p. 89.
   5. Joel O. Lubenau, "Vanadium, Stained Glass, Helpful Metal," *Western Pennsylvania History,* Winter 2011–12, p. 47.
   6. Claude S. George, Jr., *The History of Management Thought* (Englewood Cliffs, NJ: Prentice-Hall, 1968), p. 35.
   7. Peter Duncan Burchard, *George Washington Carver: For His Time and Ours* (Washington: National Park Service, 2005), p. 148.

Chapter 7

   1. Stephen Yafa, *Cotton: The Biography of a Revolutionary Fiber* (New York: Penguin, 2006), p. 235.
   2. Bruce Feller, *America's Prophet: How the Story of Moses Shaped America* (New York: HarperPerennial, 2009), p. 248.
   3. Arvarh Strickland, "The Strange Affair of the Boll Weevil: The Past as Liberator," *Agricultural History,* Vol. 68, No. 2, p. 166.
   4. Reynold M. Wik, *Henry Ford and Grassroots America* (Ann Arbor: University of Michigan Press, 1973) p. 156.
   5. Greg Grandin, *Fordlandia: The Rise and Fall of Henry Ford's Forgotten Jungle City* (New York: Picador, 2009), p. 60.
   6. John Perry, *George Washington Carver* (Nashville: Thomas Nelson, 2011), p. 39.
   7. Greg Grandon, *Fordlandia: The Rise and Fall of Henry Ford's Forgotten Jungle City* (New York: Picador, 2009), p. 278.
   8. Michael Lacey, *Ford: The Men and the Machine* (Boston: Little, Brown, 1986), p. 145.
   9. Catherine Drinker Bowen, *Yankee from Olympus* (Boston: Little, Brown, 1944), p. 79.
   10. *Detroit News,* July 12, 1916.
   11. Mark D. Hersey, *My Work Is That of Conservation: An Environmental Biography of George Washington Carver.* (Athens: University of Georgia Press, 2011), p. 162.
   12. Linda O. McMurray, *George Washington Carver: Scientist & Symbol* (New York: Oxford University Press, 1981), p. 253.
   13. Mark R. Finlay, *Growing American Rubber:*

Strategic Plants and the Politics of National Security (New Brunswick: Rutgers University Press, 2009), p. 76.

14. Paul Dickson and William Hickman, *Firestone: A Legend. A Century. A Celebration* (New York: Forbes Custom, 2000), p. 19.

15. Ronald Clark, *Edison: The Man Who Made the Future* (New York: G. P. Putman's Sons, 1977), p. 238.

Chapter 8

1. Charles Sorenson, *My Forty Years with Ford* (Detroit: Wayne State University Press, 2006), p. 172.

2. Linda O. McMurray, *George Washington Carver: Scientist & Symbol* (New York: Oxford University Press, 1981), p. 172.

3. "Urges High Peanut Tariff," *New York Times*, October 17, 1920.

4. John Perry, *George Washington Carver* (Nashville: Thomas Nelson, 2011), p. 86.

5. "Amazing Food Uses for the Lowly Peanut," *St. Louis Globe-Democrat*, April 3, 1921.

6. *Hearings Before the Committee on Ways and Means, House of Representatives on Schedule G, Agricultural Products and Provisions*, January 21, 1921, Library of Congress.

7. Gary R. Kremer, *George Washington Carver in His Own Words* (Columbia: University of Missouri Press, 1987), pp. 128–129.

8. Henry A. Wallace, "The Uniqueness of George Washington Carver," speech, October 5, 1956, Simpson College Archives.

9. Howard P. Segal *Recasting the Machine Age* (Amherst: University of Massachusetts Press 2005), p. 52.

10. "Mr. Ford Doesn't Care," *Fortune*, December 1933, p. 134.

11. Henry Ford and Samuel Crowther, *Today and Tomorrow: Special Edition of Ford's 1926 Classic* (New York: Productivity Press, 1988), p. 111.

12. Tom McCarthy, "Henry Ford: Industrial Conservationist," *Progress in Industrial Ecology: An International Journal*, Vol. 3, No. 4, 2006.

13. George Washington Carver, "The Need for Scientific Agriculture in the South," *American Monthly Review of Reviews*, Vol. 46, 1902, p. 875.

14. William Jackson, "Opening Statement, George Washington Carver National Monument," Symposium on the Life and Times of Dr. George Washington Carver, U. S. Department of Agriculture, Washington, D.C., October 7, 1999.

15. Henry Ford and Samuel Crowther, *Today and Tomorrow: Special Edition of Ford's 1926 Classic* (New York: Productivity Press, 1988), p. 184.

16. Ford R. Bryan, *Friends, Families & Forays: Scenes from the Life and Times of Henry Ford* (Detroit: Wayne State University Press, 2002), p. 151.

Chapter 9

1. Mark D. Hersey, *My Work Is That of Conservation: An Environmental Biography of George Washington Carver* (Athens: University of Georgia Press, 2011), p. 165.

2. Letter from George Washington Carver to Ernst Thompson, December 19, 1922, Tuskegee University Archives, George W. Carver Papers, Box 11.

3. Letter from George Washington Carver to L. H. Pammel, March 5, 1924, Iowa State University Archives, George Washington Carver Collection.

4. Linda O. McMurray, *George Washington Carver: Scientist & Symbol* (New York: Oxford University Press, 1981), p. 186.

5. Gary R. Kremer, *George Washington Carver in His Own Words* (Columbia: University of Missouri Press, 1987), p. 175.

6. Ibid., p. 115.

7. Howard P. Segal, *Recasting the Machine Age* (Amherst: University of Massachusetts Press, 2005), p. 168.

8. Henry Ford and Samuel Crowther, *Today and Tomorrow: Special Edition of Ford's 1926 Classic* (New York: Productivity Press, 1988), p. 239.

9. "Muscle Shoals," *Boston Ideas*, January 1 to October 1, 1927, The Henry Ford, Benson Research Center Archives, Dearborn, Michigan, linear files, Muscle Shoals.

10. Samuel Crowther, "Muscle Shoals," *McClure's Magazine*, January 1923.

11. Vivian Gunn Morris, *The Price They Paid: Desegregation in an African American Community* (New York: Columbia University Press, 2002), p. 11.

12. Littell McClung, "What Can Henry Ford Do With Muscle Shoals," *Illustrated World*, April 1922, Vol. 27, No. 2, p.185.

13. Ibid., p.186

14. Ibid., p.185

15. Speech before Congress, House of Representatives Appropriations Committee, January 17, 1923, The Henry Ford, Benson Research Center Archives, Ford Fairlane Papers, Box 5, accession number 572.

16. James Dalton, "Ford Tells What He Hopes to Do with Muscle Shoals," *Automotive Industries*, October 19, 1922, Vol. 47, No. 16, p. 1.

17. Reynold M. Wik, *Henry Ford and Grassroots America* (Ann Arbor: University of Michigan Press, 1973), p. 106.

Chapter 10

1. Gary R. Kremer, *George Washington Carver in His Own Words* (Columbia: University of Missouri Press, 1987), p. 116.
2. Linda O. McMurray, *George Washington Carver: Scientist & Symbol* (New York: Oxford University Press, 1981), p. 192.
3. Steven Watts, *The People's Tycoon: Henry Ford and the American Century* (New York: Alfred A. Knopf, 2005), p. 270.
4. Charles Sorenson, *My Forty Years with Ford* (Detroit: Wayne State University Press, 2006), p. 19.
5. Ford B. Bryan, *Beyond the Model T.* (Detroit: Wayne State University Press, 1990), p. 116.

Chapter 11

1. *The Colored Alabamian*, Vol. 3, February 1909.
2. Henry Ford and Samuel Crowther, *Today and Tomorrow: Special Edition of Ford's 1926 Classic* (New York: Productivity Press, 1988), p. 156.
3. Bruce Pietrykowski, "Fordism at Ford: Spatial Decentralization and labor Segmentation at the Ford Motor Company, 1920–1950," *Economic Geography*, Vol. 71, No. 4, October 1995.
4. Henry Ford and Samuel Crowther, *Today and Tomorrow: Special Edition of Ford's 1926 Classic* (New York: Productivity Press, 1988), p. 128.
5. *Fortune*, December 1933, p. 125.
6. *Ford News*, Vol. 19, No. 8, August 1939, p. 187.
7. Howard P. Segal, *Recasting the Machine Age* (Amherst: University of Massachusetts Press 2005), p. 45.
8. Henry Ford and Samuel Crowther, *Today and Tomorrow: Special Edition of Ford's 1926 Classic* (New York: Productivity Press, 1988), pp. 131–132.

Chapter 12

1. August W. Giebelhaus, "Farming for Fuel: The Alcohol Motor Fuel Movement of the 1930s," *Agricultural History*, Vol. 54, No. 1, January 1980, pp. 173–184.
2. Joseph Russell, "Products and the Use of Soy Beans," *Economic Geography*, Vol. 18, No. 1, January 1942.
3. Richard Leach, "The Federal Government and the Peanut Industry," *Southern Economic Journal*, Vol. 21, No. 1, July 1954, pp. 53–61.
4. H. E. Barnard, "Prospects for Industrial Uses for Farm Products," *Journal of Farm Economics*, Vol. 20, No. 1, February 1938, p. 122.
5. August Giebelhaus, "Farming for Fuel: The Alcohol Motor Fuel Movement of the 1930s," *Agricultural History*, Vol. 54, No. 1, January 1980, p. 180.
6. Lawrence Elliott, *George Washington Carver: The Man Who Overcame* (Englewood Cliffs: Prentice-Hall, 1966), p. 158.
7. Mark Finlay, *Growing American Rubber: Strategic Plants and the Politics of National Security* (New Brunswick: Rutgers University Press, 2009), p. 193.

Chapter 13

1. Malcolm W. Bingay, "Good Morning," *The Detroit Free Press*, January 7, 1943.
2. Textbook manuscript, *Botany Made Easy*, Tuskegee Institute, George Washington Carver Papers, microfilm, reel no. 47, frame 383, 1915.
3. Peter Duncan Burchard, *George Washington Carver: For His Time and Ours* (Washington: National Park Service, 2005), p. 23.
4. Ibid., p. 56
5. Henry Ford and Samuel Crowther, *Today and Tomorrow: Special Edition of Ford's 1926 Classic* (New York: Productivity Press, 1988), p. 56.
6. Booker T. Washington, *Up from Slavery: The Autobiography of Booker T. Washington* (Charleston, SC: Createspace, 2010), p. 11.
7. George Rogers and R. F. Saunders, "Henry Ford at Richmond Hill, Georgia: A Venture in Private Enterprise and Philanthropy," *The Atlanta Historical Journal*, Vol. XXIV, No. 1, Spring 1980, p. 45.
8. "The Reminiscences of Mr. H. G. Ukkleberg," The Henry Ford, Benson Research Center Archives, Oral History Section, 1951.

Chapter 14

1. James H. Cobb, "Ford and Carver Point South's Way," *Atlanta Journal*, March 17, 1940.
2. Gary R. Kremer, *George Washington Carver in His Own Words* (Columbia: University of Missouri Press, 1987), p. 161.
3. Ford Bryan, "A Prized Friendship: Henry Ford and George Washington Carver," The Henry Ford, Benson Research Center Archives, vertical file, "G. W. Carver," file 2.

Chapter 15

1. Austin Curtis Papers, Bentley Historical Library of the University of Michigan, Box 1, File 2.
2. Letter from George Washington Carver to *Peanut Journal*, Tuskegee Institute Archives microfilm, reel no. 19, frame 73.
3. Racham Holt, *George Washington Carver: An American Biography* (New York: Doubleday, Doran, 1943), p. 292.
4. Henry Ford and Samuel Crowther, *Today and Tomorrow: Special Edition of Ford's 1926 Classic* (New York: Productivity Press, 1988), p. 56.
5. Ibid., p. 89.
6. Mark D. Hersey, *My Work Is That of Conservation: An Environmental Biography of George Washington Carver* (Athens: University of Georgia Press, 2011), p. 110.
7. Henry Ford and Samuel Crowther, *Today and Tomorrow: Special Edition of Ford's 1926 Classic* (New York: Productivity Press, 1988), p. 104.

Chapter 16

1. Racham Holt, *George Washington Carver: An American Biography* (New York: Doubleday, Doran, 1943), p. 127.
2. Ibid., p. 101.
3. Gary R. Kremer, *George Washington Carver in His Own Words* (Columbia: University of Missouri Press, 1987), p. 84.
4. Steven Watts, *The People's Tycoon: Henry Ford and the American Century* (New York: Alfred A. Knopf, 2005), pp. 214–215.
5. Ford R. Bryan, *Friends, Families & Forays: Scenes from the Life and Times of Henry Ford* (Detroit: Wayne State University Press, 2002), p. 383.
6. George Washington Carver, "Nature Study and Gardening for Rural Schools," June 1910, Tuskegee Institute Archives.
7. Henry Ford and Samuel Crowther, *Today and Tomorrow: Special Edition of Ford's 1926 Classic* (New York: Productivity Press, 1988), p. 170.

Chapter 17

1. Charles Murray, *Human Accomplishment: The Pursuit of Excellence in the Arts and Science* (New York: HarperCollins, 2003), p. 428.
2. Holt Racham, *George Washington Carver: An American Biography* (New York: Doubleday, Doran, 1943), p. 257.
3. Letter from Austin Curtis to editor of *New York Times*, December 16, 1924, Austin Curtis Papers, Bentley Historical Library, University of Michigan, Box 1.

Chapter 18

1. "Better Living Through Chemurgy," *The Economist*, June 26, 2008.
2. Hugh Aldersey-Williams, *Periodic Tales: A Cultural History of the Elements from Arsenic to Zinc* (New York: HarperCollins, 2001), pp. 64–67.

Bibliography

Archives

Bentley Historical Library, University of Michigan, Austin Curtis Papers.

The Henry Ford, Benson Research Center Archives, Ford and Carver papers, and Diamond papers.

Iowa State Archives, George Washington Carver Collection.

Tuskegee University Archives, George W. Carver Papers.

Books and Articles

Aldersey-Williams, Hugh. *Periodic Tales: A Cultural History of the Elements from Arsenic to Zinc.* New York: HarperCollins, 2001.

Barnard, H. E. "Prospects for Industrial Uses for Farm Products." *Journal of Farm Economics*, Vol. 20, No. 1, Feb. 1938.

Bowen, Catherine Drinker. *Yankee from Olympus.* Boston: Little Brown, 1944.

Brinkley, Douglas. *Wheels for the World: Henry Ford, His Company, and a Century of Progress.* New York: Viking, 2003.

Bryan, Ford B. *Beyond the Model T.* Detroit: Wayne State University Press, 1990.

_____. *Clara: Mrs. Henry Ford.* Dearborn: Ford Books, 2001.

_____. *Friends, Families & Forays: Scenes from the Life and Times of Henry Ford.* Detroit: Wayne State University Press, 2002.

Burchard, Peter Duncan. *George Washington Carver: For His Time and Ours.* Washington: National Park Service, 2005.

Clark, Glenn. *The Man Who Talks with the Flowers: The Intimate Life Story of Dr. George Washington Carver.* Blacksburg: Wilder, 2011.

Clark, Ronald. *Edison: The Man Who Made the Future.* New York: G. P. Putman's Sons, 1977.

Dickson, Paul, and William Hickman, *Firestone: A Legend. A Century. A Celebration.* New York: Forbes Custom, 2000.

Elliott, Lawrence. *George Washington Carver: The Man Who Overcame.* Englewood Cliffs, NJ: Prentice-Hall, 1966.

Feller, Bruce. *America's Prophet: How the Story of Moses Shaped America.* New York: Harper-Perennial, 2009.

Ferrell, John. *Fruits of Creation: A Look at Global Sustainability as Seen Through the Eyes of George Washington Carver.* London: Macalester Park, 1995.

Finlay, Mark R. *Growing American Rubber: Strategic Plants and the Politics of National Security.* New Brunswick: Rutgers University Press, 2009.

Ford, Henry. *My Life and Work.* New York: Doubleday, Page, 1922.

_____ and Samuel Crowther. *Today and Tomorrow: Special Edition of Ford's 1926 Classic.* New York: Productivity Press, 1988.

George, Claude. *The History of Management Thought.* Englewood Cliffs, NJ: Prentice-Hall, 1968.

Giebelhaus, August. "Farming for Fuel: The Alcohol Motor Fuel Movement of the 1930s." *Agricultural History*, Vol. 54, No. 1, January 1980, pp. 173–184.

Grandin, Greg. *Fordlandia: The Rise and Fall of Henry Ford's Forgotten Jungle City.* New York: Picador, 2009.

"Henry Ford Wants Cowless Milk and Crowdless Cities." *Literary Digest*, Vol. 68, No. 9, February 26, 1921, pp. 38–42.

Hersey, Mark. *My Work Is That of Conservation: An Environmental Biography of George Washington Carver.* Athens: University of Georgia Press, 2011.

Hines, Linda. "George W. Carver and the Tuskegee Agricultural Experiment Station." *Agricultural History*, Vol. 53. No. 1.

Holt, Racham. *George Washington Carver: An American Biography.* New York: Doubleday, Doran, 1943.

Jaffe, Bernard. *Chemistry Creates a New World.* New York: Thomas Crowell, 1957.

Jones, Allen. "The South's First Black Farm Agents" *Agricultural History*, Vol. 50, No. 4, October 1976.

Bibliography

Kremer, Gary R. *George Washington Carver in His Own Words.* Columbia: University of Missouri Press, 1987.

Lacey, Michael. *Ford: The Men and the Machine.* Boston: Little, Brown, 1986.

Lane, Rose. *Henry Ford's Own Story: How a Farm Boy Rose to Power.* New York: Ellis Jones, 1917.

Leach, Richard. "The Federal Government and the Peanut Industry." *Southern Economic Journal,* Vol. 21, No. 1. July 1954, pp. 53–61.

Mackintosh, Barry. "George Washington Carver: The Making of a Myth." *The Journal of Southern History,* Vol. 42. No. 4, November 1976.

Mayberry, B. D. "The Tuskegee Movable School: A Unique Contribution to National and International Agriculture and Rural Development." *Agricultural History,* Vol. 65, No. 2, Spring 1991.

McCarthy, Tom. "Henry Ford: Industrial Conservationist." *Progress in Industrial Ecology: An International Journal,* Vol. 3. No. 4, 2006.

McClung, Littell. "What Can Henry Ford Do With Muscle Shoals." *Illustrated World,* Vol. 27, No. 2, April 1922, pp. 183–188.

McMillen, Wheeler. *The Green Frontier: Stories of Chemurgy.* New York: G. P. Putnam's Sons, 1969.

_____. *New Riches from the Soil: The Progress of Chemurgy.* New York: D. Van Nostrand, 1946.

McMurray, Linda O. *George Washington Carver: Scientist & Symbol.* New York: Oxford University Press, 1981.

Minnich, Harvey. *William Holmes McGuffey and His Readers.* New York: American Book Company, 1936.

Morris, Vivian Gunn. *The Price They Paid: Desegregation in an African American Community.* New York: Columbia University Press, 2002.

Murray, Charles. *Human Accomplishment: The Pursuit of Excellence in the Arts and Science.* New York: HarperCollins, 2003.

Olson, Sidney. *Young Henry Ford: A Pictorial of the Firs Forty Years.* Detroit: Wayne State University Press, 1997.

Perry, John. *George Washington Carver.* Nashville: Thomas Nelson, 2011.

Pietrykowski, Bruce. "Fordism at Ford: Spatial Decentralization and labor Segmentation at the Ford Motor Company, 1920–1950." *Economic Geography,* Vol. 71, No. 4, October 1995.

Proceedings of the Dearborn Conference on Agriculture, Industry, and Science. Dearborn, Michigan, May 7 and 8, 1935.

Pursell, Carroll. "The Farm Chemurgic Movement Council and the United States Department of Agriculture." *History of Science and Society,* Vol. 60, No. 3, Autumn 1969.

"The Reminiscences of Mr. H. G. Ukkleberg." Benson Ford Archives, Oral History Section, 1951.

Rogers, George, and R. F. Saunders. "Henry Ford at Richmond Hill, Georgia: A Venture in Private Enterprise and Philanthropy." *The Atlanta Historical Journal,* Vol. XXIV, No. 1, Spring 1980, pp. 43–51.

Segal, Howard P. *Recasting the Machine Age.* Amherst: University of Massachusetts Press, 2005.

Shurtleff, William. *Henry Ford and His Researchers' Work with Soybeans, Soy Foods, and Chemurgy.* Lafayette, CA: Soy Foods Center, 1997.

Skrabec, Quentin R. *William McGuffey: Mentor to American Industry.* New York: Algora, 2009.

Sorensen, Charles E. *My Forty Years with Ford.* Detroit: Wayne State University Press, 2006.

Strickland, Arvarh. "The Strange Affair of the Boll Weevil: The Past as Liberator." *Agricultural History,* Vol. 68, No. 2.

Third Dearborn Conference. *Farm Chemurgic Journal,* Vol. 1, No. 1, September 17, 1938.

Washington, Booker T. "A Retrospect and Prospect." *The North American Review,* Vol. 182, No. 593, April 1906.

_____. *Up from Slavery: The Autobiography of Booker T. Washington.* Charleston, SC: Createspace, 2010.

Watts, Steven. *The People's Tycoon: Henry Ford and the American Century.* New York: Alfred A. Knopf, 2005.

Wellman, Paul. "Friends of Old Days in Kansas Saw Budding Genius of Negro Scientist." *Kansas City Star,* September 9, 1942.

Wik, Reynold M. *Henry Ford and Grass-roots America.* Ann Arbor: University of Michigan Press, 1973.

_____. "Henry Ford's Science and Technology for Rural America." *Technology and Culture,* Vol. 3, No. 3, Summer 1962.

Yafa, Stephen. *Cotton: The Biography of a Revolutionary Fiber.* New York: Penguin, 2006.

Index

acorns 50, 58
Agrol 146–147
alcohol based fuel 8, 13–14, 19, 86–87, 143–148, 194–195
alcohol based rubber 152–154, 165
aluminum 116, 197
animal feeds 58–59
artificial cloth 102
artificial leather 102, 196
artificial wool 102, 147, 170
assembly line 73–76, 102–104
Atchison alcohol plant 146
Atlanta 107–108
Avery, Clarence 70, 75

Baekeland, Leo 147–148
Bakelite 148
bamboo 153
Barnard, H.E. 145–146, 152
Barthel, Oliver 41
Bell, Alexander 190
Berry, Martha 183
Berry School 183–185
Bessemer, Henry 190
boll weevil 57, 78–80
Botsford Inn 105
Bourbon Democrats 108–109
bread making 103–104
Bridgeforth, George 56
Budd, Etta 31–32
Burbank, Luther 5–7, 12, 16, 32, 40, 89, 92, 187
Burroughs, John 5–7, 16, 75, 89, 90
by-product use 100–103, 148, 175

cabbages 77
Campbell, Thomas 68–72
Carnegie, Andrew 44, 47, 58, 188
carrots 83

Carver, George Washington: boyhood 21–26, 29–32; conservation 76; early career 36–61, 64–67; friendship with Ford 8–10, 17–20, 145–153, 154–163; general 1, 5, 6, 7, 13–15, 19; later years 119–122, 136–140, 171–174, 178–180; mid-career 77–84, 93–99, 106–113; Moses figure 78–79; paintings 32–33, 37–38; research approach 97, 137–140, 171–173, 186–193; training and teaching by Carver 64–68, 178–180; work with polio 137–140
Carver, Jim 21–24
Carver, Moses 21–22
Carver, Susan 22–23
Carver Nutritional Lab 168–170
Carver Products Company 107
Carvoline 109
celluloid 147–148
Chautauqua movement 123
Chemurgic Farm Council 145–147
Chemurgy 5, 8, 12–16, 19, 140–154, 191, 193–194
Cherry Hill Plant 183
Chicago World's Fair of 1893 32–33, 36–39
Chicago World's Fair of 1933 125
China 94–95
Clay, Henry 12
Clays 30–31, 51, 69, 107–108, 113, 119
clover 50
Commission on Interracial Cooperation 110–111
composites 148–150

Coolidge, Calvin 16, 88, 96
corn 143–144
cotton 46, 55–57, 79–80, 112–114, 163, 165–166, 190
Couzens, James 62, 70
cowpeas 50–54, 164
Creative Chemistry 191
cresosote 120
Currie, Madame 11
Curtis, Austin 8, 109, 136–140, 169

Dearborn (Michigan) 11, 15–16, 21, 111, 154
Declaration of Dependence Upon the Soil 11–16
Detroit 34, 85–86, 103–104
Detroit Dry Dock Company 28
Detroit, Toledo & Ironton Railroad 104
Diamond, Holton 169
Diamond Grove 29–30, 197
Dodge Brothers 70
Dow Chemical Company 13, 141–142, 154, 193
Dundee 6
dyes 30–31, 51, 53, 55, 107, 112

Earle, Franklin 58
Edison, Thomas 5, 7, 11, 13, 16–19, 42, 61, 88–92, 114–115, 121–123, 160–161, 187–189
Einstein, Albert 11
Eisenhower, Dwight 16, 88
electric cars 61, 91–92
Emerson, Waldo 26
ethanol 8, 10, 14, 143–144, 194–195

Fair Lane (Ford estate) 83–84
fertilizers 53–54, 56, 66–67,

Index

77, 114, 117–118, 160, 171, 173–178
Firestone, Harvey 5, 7, 11, 13, 16, 19, 70, 79, 87–88, 90–91, 155
Firestone, Martha 88
Fisher, Fred 70
flax 102, 158–160
Ford, Bill 88
Ford, Edsel 104
Ford, Henry: boyhood 21–26; conservation 27, 75–76, 84–85, 100–106, 195–196; early career 28, 33–35, 41–42, 61–63, 70–75; friendship with Carver 8–10, 17–20, 145–153, 154–163; general 1, 5, 7, 13–15, 51, 59–60; hospitals 121, 125; later years 122–136, 174–178; management style 83–85, 100, 127–130, 186–193; mid-career 84–92, 99–105, 113–119; racing 62–63; schools 121–122, 178–184; work with disabled 183–184
Ford, Henry II 135
Ford, Mrs. Henry (Clara Bryant) 33–34, 70, 85–86, 91, 105, 121, 123, 162, 180, 183–185
Ford, William 32–33
Ford Apprentice School 182–183
Ford Museum and Greenfield Village 2–3, 7–9, 11, 14, 18, 25–27, 41–42, 122–127, 167–170, 180–182
Ford Service School 183–184
Ford tractors 86–87
Fordism 104, 130–132
Fordite 102
Fordney-McCumber Tariff Bill 95–97, 142
Fort Myers (Fla.) 88, 92, 103, 122
frog leaf spot 55

Garvan, Francis 141–145
gasoline 14, 42, 61–62, 152–153
glues 53, 163
Green Island (NY) 90
Gross Pointe (Mich.) 61

Hale, William 13, 141–145, 154
Harding, Warren 7, 88, 91
Heinz, J.H. 62–63
hemp 12, 122, 148–149, 155–157, 196; paper 157
Henry Ford & Son 87
Henry Ford Motor Company 62–63
Henry Ford Trade School 181–183
Highland Park Plant 74–75, 100, 103
Hoover, Herbert 7, 11, 15, 88, 91, 122–123
Huff, "Spider" Ed 41
hybridization 57, 89, 187

Independence Hall 2, 11, 122–123
Iowa State College 30–32, 39–41
Ireland factory 113–114
ISO 14000 195

Jefferson, Thomas 12
Jesup, Morris 68
Jesup Wagon 67–69

Kahn, Albert 75, 133
Kansas 23–24, 30–31
Kellogg Company 81–82
Kettering, Charles 143
King, Charles 41–42
Ku Klux Klan 111

Lean Manufacturing 8–9, 100, 102–104, 174–178
Leland, Henry 103
Lewis, Charles 74
Liebig, Justus Von 54
Lincoln, Abe 12, 93–94
Lincoln Motor Company 103
linoleum 81
linseed oil 159
Lovett, Benjamin 105

Malcolmson, Alexander 62–63
Martha Mary Chapel 11, 124, 180, 185
McGuffey, William 11, 25–26, 60, 66, 180–181
McGuffey Readers 23–27, 60, 141, 180–181, 185, 197–198

McKinley, William 39, 57, 68, 114
McMillen, Wheeler 142–145
Model N 69–72, 74
Model T 5, 9, 14, 63, 69, 71–74, 90–92, 100, 143, 179, 193
Montgomery (AL) 58–59, 94
Moton R.R. 116
mulberry trees 55
Muscle Shoals 6, 103, 114–118, 163
mycology 55, 57–58, 187–188

NAACP 109–110
New Deal 7, 12, 15–16, 118, 144–146
Norris, George 117
Northville 6, 103, 131–132
nuts 171
nylon 152

okra 51, 196
Orphan Home of Detroit 180
Osage oranges 166

paints 59
Pammel, Louis 40, 192
Peano oil 109
peanut milk 81–82, 96
peanuts 6, 15, 53, 80–82, 93–97, 112–113, 145, 163–166, 196–197
Penol 109, 120–121
persimmons 166
Pierce, John 68–69
Piquette Plant 74–75
plastics 147–150, 194–196
Plymouth 6
Prohibition 87–88
protectionism 93–97
Pure Food and Drug Act 1906 58

quadricycle 41–42, 61

racism 18
ramie 152–153
rayon 147
recycling 100–107, 171–178
Reichhold Chemicals 152
reinforced plastics 195–196
Richmond Hill Plantation 159–162, 165–166, 184–185
River Rouge Plant 86, 93,

Index

100, 102–103, 113, 172–178, 182–183
road building materials 112
Roberts, Peter 179
Rockefeller, John D. 44, 59, 189
Rogers, Will 11
Roosevelt, Franklin 15, 39, 143, 193
Roosevelt, Teddy 47, 49, 68, 114
Roosevelt Medal 192
rubber 13, 18–19, 90–92, 111, 147, 150–152, 160, 168, 197
rubber plantations (Fordlandia) 128–132, 197
Ruddiman, Edsel 25, 87, 125
Russia 86

saline 6
Schwab, Charles 181
Sibley, Harper 13
Simpson College 31, 192
Sorensen, Charles 73–74, 86, 103, 124
soybeans 6, 15, 55, 125–127, 144, 149–151; car 148–150, 195–197; foam 193–194; milk 125; plastics 125–126, 144, 170, 195–197
Sparks, E.E. 97

Standard Oil 143, 151
Stine, Charles 13
Strauss, Fredrick 28, 42
sweet potatoes 6, 19, 50–51, 54–55, 69, 82–84, 87–88, 119–120, 149, 163–165, 190
synthetic rubber 151–153, 165

Taft, William 7, 39–40
tariffs 93–97
Thompson, Ernst 107–108, 137
Toyota 2, 5, 9, 177
Truman, Harry 26, 39
Tuskegee Institute 5, 18, 40–42, 43–63, 65–68, 111–112, 116, 137–140, 170–173, 180–182

Ukkelberg, Harry 161–162
United Peanut Growers Association 94

vagabonds 88–89, 90–91
vanadium steel 71–73
vegetable oil uses 150–153
village industries 6, 103, 126–140, 163

Wallace, Henry A. 16, 39, 97, 178

Wallace, Henry C. 39–40
Wandersee, John 75
Washington, Booker T. 44–45, 48–50, 58–60, 65–68, 77–78, 178–180
Ways, Georgia 129–130, 152, 183–185
Webster, Noah 11
Weeks-McLean Bird Bill 27, 75–76
Westinghouse Company 32–33
Whig Party 140–141
Willow Run 6
Wills, Harold 70–73
Willys-Overland Motors 70
Wilson, James 34, 40, 47, 49, 55, 68, 95–96
Wilson, Woodrow 96
Wilson dam 115
Winton, Alexander 61–63
wood alcohol 87–88, 101
wood stains 69
World War I 82–83, 88, 114, 143, 163, 190
World War II 163–170, 190

YMCA 59, 110–110, 179–180
Ypsilanti 6

www.ingramcontent.com/pod-product-compliance
Lightning Source LLC
Chambersburg PA
CBHW081556300426
44116CB00015B/2898